Also by Arthur Upgren

Weather: How It Works and Why It Matters
(coauthored with Jurgen Stock)

Night Has a Thousand Eyes:
A Naked-Eye Guide to the Sky,
Its Science and Lore

✻

The Turtle and the Stars

The Turtle and the Stars

OBSERVATIONS OF AN EARTHBOUND ASTRONOMER

Arthur Upgren

AN OWL BOOK

Henry Holt and Company · New York

Henry Holt and Company, LLC
Publishers since 1866
115 West 18th Street
New York, New York 10011

Henry Holt® is a registered trademark of
Henry Holt and Company, LLC.

Library of Congress Cataloging-in-Publication Data

Upgren, Arthur R.
  The turtle and the stars : observations of an earthbound
astronomer / Arthur Upgren.
    p. cm.
  Includes bibliographical references and index.
  ISBN 0-8050-7290-X (pbk.)
    1. Astronomy—Popular works. I. Title.

QB44.3.U64 2002
520—dc21   2001052741

Henry Holt books are available for special promotions and
premiums. For details contact: Director, Special Markets.

First published in hardcover in 2002 by Times Books

First Owl Books Edition 2003

An Owl Book

Designed by Victoria Hartman

Printed in the United States of America

1 3 5 7 9 10 8 6 4 2

To Joan and Amy,

who shared my encounter

with the mother leatherback turtle

in the defining moment of this book

———————

And to all who cherish

the sight of the night sky filled

with stars and the Milky Way

✳

Contents

✳

Preface

The important thing is not to stop questioning.
Albert Einstein

Perception and reality may contradict one another in many ways as we look up at the sky. This can be readily illustrated by perhaps the most widely known example of the difference, the obvious impression that our world is a big flat stationary place, and those little stars up there are tiny and do all the flitting about that we see them do. And besides, if the Earth were round, how do the Australians keep from falling off? I recall being perplexed as a young child, reading that the Sun is bigger than the Earth. That bright little thing up there larger than this whole world with its cities and farms and seas? I didn't buy that until my father, with the help of an encyclopedia, confirmed the truth of it. Even then it took me a while to be comfortable with it.

Things far away look small; I both knew that and didn't know it, probably because I never thought it through. My earliest perceptions of this truth were incomplete, even if not inaccurate. Since that time, I have learned how facts can be arranged and explained to provide an understanding, a synergy that exceeds their sum as facts alone. That is what I have devoted my life to as a teacher, and that is what I mean to convey here.

No discipline combines the large and the small, the nearby and the far off, the great numbers and the enchanting night sky beauty as does astronomy. Often proclaimed our first science, the inherent simplicity of the Sun, the Moon, and five bright planets moving about nearly repetitively against a backdrop of all those stars and star-figures appealed to the curious as did no other.

In this book we will look at things large and small that are seen as a part of the sky, such as towers and bridges, that can illuminate scientific principles in an imaginative way. Relationships such as these do change perceptions, generally toward a fuller understanding of a theory or principle. Truth, reality, may be stranger than fiction or it may not, but it is frequently more rewarding in its comprehension.

When perception on the part of the public clashes with reality, the scientist has a responsibility to point out how and why the popular view is inaccurate or incorrect. For example, many readers subscribe to the belief that Nostradamus, the sixteenth-century physician and astrologer, was able to foretell the future. Who today would not be startled to read the following quatrain, which he wrote before 1568 as a prediction for the middle of our twentieth century:

> Bestes farouches de faim fleuvis tranner,
> Plus part du champ encontre Hister sera.
> En caige de fer le grand fera treisner,
> Quand rien enfant de Germain observera.

> Beasts wild with hunger will cross the rivers,
> the greater part of the battlefield will be against Hister.
> He will drag the leader in a cage of iron,
> when the child of Germany observes no law.

The close resemblance to the Third Reich and its Nazi bestiality, which observed no law but its own, strikes many as remarkable. This singular forecast with its Hister/Hitler proximity is one among hundreds of others, most of which have little or no predictive value or association with subsequent events. Most scientists put this down to a lucky guess, one hit among many misses, near and far. Woody Allen goes further, mocking these predictions with one of his own. He predicted that "two countries will go to war, but only one will win." With this broad statement, he is far more often right than Nostradamus. Believers in the predictions of Nostradamus or Edgar Cayce fail to take those many misses into account, but recall only the occasional hit; such neglect for all of the available evidence can lead promptly and erroneously to a belief in the paranormal.

Very few among us, believers and skeptics alike, can explain whether or why an egg can or cannot be balanced on its end only during the times of the equinoxes. Yet the underlying principle here is among the foundations of the scientific revolution that led Kepler and Newton to shape our cosmos. A failure to comprehend it opens the way to further acceptance of pseudoscience.

These are among many mistaken impressions for which a supernatural explanation seems entirely reasonable. But science is an unforgiving taskmaster and requires that the response be consistent with its rules. Over thousands of years, the human race has built an edifice of rules that are mutually consistent.

＊

LET US BEGIN with some of the sights of astronomy that form the underlying appeal of the sky and its wonders. These are the spectacles, rare and commonplace, that bring one to appreciate the world, the larger world, and its sky.

From prehistoric times, people have tried to model the world and the sky and portray it all in terms and images they can comprehend. The sky appears deceivingly simple—just seven objects: the Sun, the Moon, and five bright planets in motion against a background of stars fixed in place and moving only to reflect the rotation of the Earth, once every 24 hours. Yet people have known or guessed that seeing the cosmos as we do from one point within, we must conjure a total picture of it as it might appear from afar to bring order and sense to it.

Many different attempts have been made to do this. One of the best known is from ancient India, where the world was allegedly pictured as a hemisphere mounted on the backs of four huge elephants that stood on a giant turtle, which in turn stood on a coiled snake. Might this have been a multistage cosmology, in that the elephants answered the need to know what held up the world, until the inevitable next questions: What supports the elephants, and later the turtle, and still later the snake? Below the snake there is mud all the way down, a response that sounds like a parent's weary answer to a curious child. Not for another millennium did models develop that were more successful at "preserving the phenomena," a phrase from the Hellenic times of Plato and Aristotle that called for theory to conform to observation and established fact. Thus began the enterprise we call science.

The nonempirical sciences—mathematics and logic—develop through propositions proved without reference to observation or other empirical evidence. In contrast, the so-called empirical sciences "seek to explore, to describe, to explain, and to predict the occurrences in the world we live in . . . the high prestige that science enjoys today is no doubt attributable in large measure to the striking successes and the rapidly expanding reach of its applications." When these words were written and published in 1966, by Carl Hempel in *Philosophy of Natural Science,* the high prestige he cites was greater and more evident than it is today, more than 30 years afterward. Later on in this book, I shall expand on this theme; suffice it here to take note of it. The chapters reveal in ever widening percipience the wonder and the charm of reality in our profound world and cosmos, how phenomena from around the world can enhance an understanding of it, and how knowledge can be used to clarify or correct a number of common and widely believed misconceptions. Astronomers live with a larger mix of pseudoscience than most other scientists; astrology and UFOs, creationism and alien landings, light pollution brightening the night sky and the Full Moon affecting our sanity are based upon unsound and sometimes untrue foundations, all of which preclude that special joy of discovery that comes only with understanding.

In the book *A Land,* written half a century ago, and one that deserves wider recognition than it now enjoys, Jacquetta Hawkes evokes an image of an entity of Britain in which past, present, nature, man, and art appear all of a piece, and notes that "the nature of this beauty cannot be stated, for it remains always just beyond the threshold of intellectual comprehension.

It can only be shown as a blurred reflection through hints coming from many directions but always falling short of their objective."

This is an attempt to do the same for the sky. Many are the thoughts that pass through our minds when we look at the dark moonless night sky. This great empty dome, strewn with stars and riven by the paleness of the Milky Way stretching from one horizon to the other, has always been associated with heaven, God's abode.

We, most of us, no longer subscribe to the claustrophobia that C. S. Lewis discusses in *The Discarded Image,* his introduction to medieval life and literature. The ordered medieval cosmos he describes was too orderly for our post-Copernican tastes; we know of uncertainties such as the big bang and the weird peculiar objects thrown up by it, which now reveal themselves to the Hubble space telescope, and we are aware that we still don't know the size or age of our universe.

I sometimes wish that I could once again see the sky as I did as a young child before a lifetime of study provided some of the answers. What if the crescent Moon really were a silvery thing only several feet in size upon which I could perch and make a wish? Might not the stars be displaced Christmas-tree lights twinkling up there, instead of seething spheres of overheated hydrogen gas? Can this childlike but not childish wonder ever be recaptured? Were we somehow closer to the stars in the days before the all-pervasive baleful orange glare blotted out the wondrous Milky Way? They haven't gone away but we have; we have left them, cluttering our skies with light and filth. Perhaps we may still return and when we do, they'll be there to welcome us home.

*

Introduction
Dark Skies: A Right, Not a Privilege

> And God said, Let there be light: and there was light.
> And God saw the light, that it was good:
> and God divided the light from the darkness.
>
> Genesis 1:3–4

One does not have to go far in the Bible to find light equated with goodness. Most, if not all, of our languages are steeped in the equivalence, and its obverse, that darkness is a manifestation of evil. The light at the end of the tunnel, a sunny disposition, a shady deal, a bright child, a dark thought are all expressions to help us to reaffirm the association of the good guys with the white hats and the bad guys with the black hats.

The list is endless; any thesaurus (e.g., *Roget's Twenty-first Century Thesaurus*) shows for light a number of synonyms such as: brightness, radiance, sparkle, splendor, and many other feel-good words. As a verb, light matches animate and shine. Darkness can bring us gloom, murk, and dead of night; or ignorant, mysterious, and arcane; or drab and dull. Then we can become further enlightened by proceeding on to bad, corrupt, evil, satanic, cheerless, ominous, brooding, sinister, scowling, threatening, and even Stygian.

Where should darkness occur in any form except perhaps in a photo lab or a haunted house at Disney World? Those of us who seek to trim lumens and wattage face millennia of associations of these kinds. Where do we get off preaching the virtues of the dark? Should we leave that insidious underworld to Edgar Allan Poe and H. P. Lovecraft, and even Fyodor Dostoyevsky at one remove; never mind that Herman Melville's great evil whale with a twisted jaw was white. Indeed it was Lovecraft who commented in the essay "Supernatural Horror in Literature" that "the oldest and strongest emotion of mankind is fear, and the oldest and strongest kind of fear is fear of the unknown." Darkness is one of the fears he traces, one felt by many to harbor the unknown.

Why indeed, we read further in the Good Book, did God go on to make a firmament and call it Heaven? He divided the day from the night, to be governed by the Sun, and the Moon and stars, respectively. The nyctophobes among us might suppose that night was the creation of Satan, not God. God could just as easily have located us in a double or multiple star system, between two or more great suns. In *Nightfall,* the novella that brought him his first literary success at the age of twenty-one, Isaac Asimov posed a planet in a sextuple star system, which bathed it in perpetual daylight but for once in a thousand years when the calamity of night struck briefly, and out of fear and panic the civilization tore itself apart. We might have developed in such a system, with eternal daylight, and the circadian rhythm of light and darkness to which the Earth's children have evolved and adapted might never have occurred.

But we earthlings have always experienced alternating daily rhythms of daylight and darkness, extremes in light and dark. Even 570 million years ago, when the single-celled life first and suddenly burst forth into large multicelled life-forms—fish, trilobites, and other marine life—the length of the day was just a few hours shorter than it is today. All creatures since that time have known day and night about as we know it now.

The difference in light intensity between the daytime sky and the sky on a moonless night is more than 10 million to one; no other stimuli come remotely close to subjecting other sensory input to such wide-ranging yet normal variations as day and night do to our eyesight. There are no hours-long periods of noise alternating with quiet interludes of such difference, nor periodic variations affecting smell, taste, or feel. Nor does our natural world shuttle between wide extremes in air pollution or any other kind of atmospheric dislocation.

Day and night are and always will be a normal and fundamental part of our total experience. To shun such an indelible heritage as the rhythm of night and day for one of eternal brightness is to evade the nature that has been created for us.

*

FOR MOST OF my life, I have gazed upon the natural night sky with rapture and curiosity. Early on in my lifelong love affair with the sky, I not only questioned the size of the Sun, but I was also confused when my mother told me that there are people living down below us somewhere who get up in the morning when we go to bed, and who go to bed when we rise. I pictured them living in basements or underground garages, and needed a few more details before I could get a firmer grasp on the whole global picture. My formative years included both urban and rural living, and I knew early on that skies are perceived differently in the two arrangements. The city sky is as full of weather changes as any other, but who really notices or cares unless it is windy, wet, or snowing and we are directly affected? At night, the city sky revealed the Moon and a star or two between the trees and buildings, but in the country with an unbroken celestial hemisphere, alterations in cloud or weather were immediately noticeable, and at night the Milky Way banded the starry sky from one horizon to the other. Glows of nearby towns were just visible along the horizon, and mapping them onto an upside-down likeness of rural Wisconsin where I lived became a source of great interest to me.

In the nearby towns and in the residential areas of the city, the streets were delineated by overhead incandescent bulbs perched on extensions from lampposts reaching out over the centers of the streets. The wind blew them around, and they looked feeble and forlorn in the long nights of wintertime when the trees were bare and the land was buried under interminable ice and snow. The lights seemed almost friendly against such a harsh landscape.

I didn't appreciate how lucky I was then, to have dark skies to ponder. Not until later did the harsh blue glare of the mercury vapor lamps spread from the cities to the suburbs and countryside. Even then one could behold the night sky in partial glory. For me, the nocturnal world was obliterated in the 1970s when the dreary orange contagion of high-pressure sodium bulbs spread into the landscape.

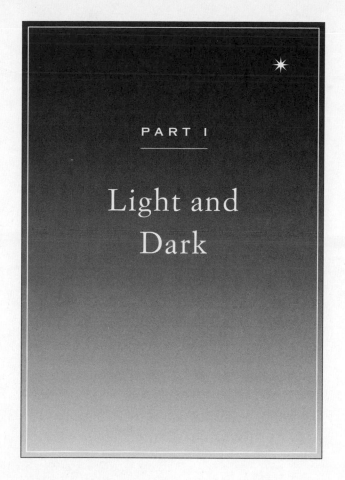

PART I

Light and Dark

1

✳

The Turtle and the Stars

In January, a few years ago, on a beach along the western shore of Costa Rica, my wife, daughter, and I had the privilege of witnessing a giant leatherback turtle emerge from the waves onto the land. By the light of a small red flashlight and the stars, we saw her dig a hole in the sand with her rear flippers, deposit her eggs into it, and head back to the sea.

I have long known that sea turtles did this, but I was not at all prepared for the visual impact of the scene. In the western sky that moonless night, high above the water, Venus shone among the stars with a brilliance rarely seen outside the tropics. Supplemented by the lambent light from its reflection upon the waves, it provided illumination sufficient for our dark-adapted eyes to view the seascape and the massive creatures moving upon it.

The appearance of the sky is almost as familiar to me as is my face in the mirror. Yet never had I seen the evening star cast more light than it did on that night. I don't think I had ever before been a spectator at a terrestrial event that was not illuminated by the Sun or by the Moon or by artificial light. This brightest of all starlike objects cast shadows and provided direction to the night around us much as does the Moon when it is visible.

We had arranged this evening tour in the course of a leisurely visit to one of the many lovely seaside resort hotels facing the Pacific surf. During the previous year rapidly developing cataracts in both eyes left me nearly blind, but just before this venture one lens had been successfully replaced with an implant. We had gone to this small peaceful land for a vacation and to see the varied topography for which it is justly famous, but for me it became also a celebration of regained eyesight. Hillside coffee fields, mountain lakes, and tropical rain forests above the clouds await the visitor, and one spectacular tour features an active volcano throwing out lava and boulders that produce a reddish glow in the twilight sky.

But the indisputable highlight of this trip came upon us unexpectedly on the memorable night of the turtle watch. On that darkened strand small groups of tourists followed guides armed only with red flashlights; our group consisted of about a dozen following a guide in the near blackness. I saw once again those many bright stars that render the tropical sky so much more spectacular than our northern one at home. As we proceeded, our night vision improved and soon we could easily discern what appeared to be large boulders strewn along the beach. We could see that a few of them moved slowly toward or away from the sea. Clearly they were giant turtles in the process of sustaining their species. A sense of the importance of darkness to this pageant became evident to all of us when someone briefly shone a flashlight unshielded by a red filter on one of the turtles. She promptly reversed her landward motion and disappeared beneath the sea, abandoning her mission. We grouped behind one of the plodding masses and followed her by the light of Venus. Under the starry sky the giant 1,200-pound leatherback sea turtle emerged without haste from the water onto the beach and made her ponderous way landward to lay her eggs in a rite of fertility reaching back to the time of the dinosaurs. No outdoor lighting marred the scene or scared her off. Regulations forbade lights on the beach except for low-power red ones near the few homes at some distance from us.

The turtle had begun to dig a hole in the sand with her back flippers. Only now that this ritual was under way was it safe for us to gather around closely without disturbing her. I was struck by the way her species and mine could almost touch with no apparent apprehension on either side. I felt no fear of this majestic being who outweighed me by half a ton. I felt only a sense of humility beside so successful a survivor.

Slowly and steadily she worked. Each rear flipper in turn scooped sand out, deepening the hole. As we quietly watched her labors, I was struck with the mechanistic manner in which her flippers moved. They were the fitful motions of the farm and factory machinery of my childhood and the belts, gears, and pulleys that produced a slow jolting motion with the regular timing of a pendulum. They seemed to move with none of the arhythmicity associated with living things. Slowly and methodically each flipper in turn went through its preset motions throughout the entire ritual and she took perhaps an hour to complete the hole to her satisfaction.

At last this gentle giant came to the end of her orderly scooping. Some primal perception signaled to her that the hole underneath her tail was deep enough, being somewhat over 2 feet in depth. Venus had lowered in the west, and in the east a spectacular array of bright objects had risen. Mars and the two brightest stars in our night sky, Sirius and Canopus, straddled the eastern and southeastern heavens. This luminous triad is formed of three of the four next brightest objects in the entire sky after Venus; only Jupiter was missing from the all-star cast. Yet our sister planet outshone each of them by ten to twenty times and still dominated the heavens.

Our turtle deposited several dozen eggs in the hole. Each was round and about the size of a tennis ball. She then pushed the piles of sand back on top of the eggs as laboriously as she had dug them out. After a pause, this was followed by a kind of rocking motion of her body back and forth in a slow rhythmic dance. We were told that its purpose was to camouflage the location of her future offspring. Clearly in no hurry, she smoothed the surface under herself, then paused again as if in satisfaction of her work before slowly turning around and making for the ocean. Largest of all turtles, she would repeat this procedure seven to ten times over a year or more and then take about three years off, according to our guide, only to go through the entire cycle afterward once again. Since her own birth, this would be her only activity on land.

On her way back to the water she was momentarily confused by another unreddened flashlight beam shining her way, but when it passed she resumed her course into the sea and soon disappeared beneath the waves. The beach remained dark; no artificial lights shone there as they would on most American beaches. The sky overhead remained dominated by Venus in the west and Mars in the east. These two most earth-like of all known celestial bodies have defied all attempts to detect life on them, and I was reminded of the uniqueness of oceans and of known life on this world. The Milky Way arched vividly overhead from one horizon to the other. This galaxy of ours contains several hundred billion stars, but are any of them accompanied by a planet bearing life? The turtle and the stars together proclaimed the dark as part of our singular planet's heritage. Her kind has depended on darkness since Tyrannosaurus rex menaced the landscape. But unlike that monster, this creature's kin managed to survive the fiery devastation of 65 million years ago when a huge comet or asteroid smashed into the Earth.

For a far longer time than Homo sapiens has been around, leatherback sea turtles and other similar reptiles have evolved to adapt to the circadian rhythm of light and dark. I hope that on this shore and many others as well, my descendants will not find a world too bright and too crowded to behold this turtle's descendants repeating their act of fertility in times to come. Current laws in Costa Rica restrict outdoor lighting in order to preserve the habitat for sea turtles to lay their eggs. Similar legislation is needed on beaches in the United States and elsewhere in order to retain for all of us the grandeur of that unforgettable night of the turtle and the stars.

2

✳

Nightfall

A Saturday afternoon in November was approaching the time of twilight, and the vast tract of unenclosed wild known as Egdon Heath embrowned itself moment by moment. Overhead the hollow stretch of whitish cloud shutting out the sky was as a tent which had the whole heath for a floor.

Thus did Thomas Hardy open his classic novel *The Return of the Native,* published more than a century ago. Hardy has here touched upon a magic moment between dark and daylight that forms a permutation within the span of sensitivity in that remarkable device, the eye. Sometime in the distant past about half a billion years ago, our premammalian ancestors evolved to form light-sensitive groups of cells with a transparent lens in front of them. Over time each group of cells became a retina around which the modern eye developed. The human eye, no doubt through a prolonged evolutionary process, spans the stretch of spectrum known as visible light. In a refinement of this evolutionary coup, the maximum sensitivity of human eyesight is in the spectral region we call yellow light at exactly the point where the Sun shines its brightest. Our eyes are sensitive to as much as 92 percent of all sunlight.

Hardy goes on to say, "The heaven being spread with this pallid screen and the earth with the darkest vegetation, their meeting-line at the horizon was clearly marked. In such contrast the heath wore the appearance of an installment of night which had taken up its place before its astronomical hour was come: darkness had to a great extent arrived hereon, while day stood distinct in the sky. Looking upwards, a furze-cutter would have been inclined to continue work; looking down, he would have decided to finish his faggot and go home."

This is the time of dusk when a ball thrown aloft can still be seen easily against the still-bright sky and caught but is lost in obscurity when viewed against trees or the ground. The light difference is phenomenal between an open field widely exposed to the open sky above and an eternally shady glen appearing already in a state of gloom when trolls and such other nocturnal folk begin to peer out upon their world. Count Dracula, were he to exist, would about now be leaving his musty casket to rule his sunless domain. It is near the moment of civil twilight, defined as that point when the Sun is 6° (twelve of its own apparent diameters) below the true western horizon. (This is about the moment that headlights should be turned on while driving.)

At the equator, civil twilight is reached 22 minutes after the moment of sunset; it would be 24 minutes if atmospheric refraction did not appear to lift the Sun such that it sets 2 minutes later (and rises that much earlier). At 41°N, the latitude of New York, civil twilight falls between 27 and 34 minutes after sunset, and in London, at 51½°N latitude, it follows sunset by 33 to 48 minutes. In both cases the amount is longer in winter and summer, and shorter in the spring and fall. Laws in most states call for headlights to be turned on from 30 minutes after sunset until 30 minutes before sunrise, not a bad approximation of the length of civil twilight in American midlatitudes.

Light levels are difficult to evaluate, partly because they cannot be reliably committed to memory, even over a time span of a few seconds. The complexity of light in its many guises is presented in an appendix; suffice it to be said here that we can think of light in brightness units of foot-candles. A foot-candle is roughly the brightness of an ordinary candle as seen from a distance of 1 foot. This is sufficient to see curbs and potholes and other potential obstacles when walking at night. In contrast, the Full Moon at its brightest produces only one fiftieth (2 percent) of a foot-candle.

The daytime sky is brighter, very much brighter. The Sun provides as much light as half a million full moons, more than could be crowded into the entire celestial sphere, even though the two appear the same size in the sky. On a clear day when the Sun is high, it shines with about 10,000 foot-candles; as we all know, direct sunlight is very bright indeed. What is not at all well recognized is the falloff of light intensity as the Sun approaches the horizon. Right at sunset, the light level is reduced by about fifteen to twenty times, to a light level of only about 500 foot-candles, and by the moment of civil twilight it has been reduced to a dusky 1.5 foot-candles. Our eyes are so adaptable to changing light intensity that we scarcely notice the change between noon and sunset, but a camera is not fooled so easily if only because film does not have that enormous range. The ratio between full daylight and the easily visible light from a candle a mile away is greater than 100 million to one, a simply incredible range.

All of us are aware that the sky seems to be nearly as bright and as blue at sunset as it was earlier in the afternoon, yet the camera shows the true light ratio through a badly underexposed photograph at sunset at a proper setting for full daylight. A photograph taken under the same conditions at the time of civil twilight, 36 minutes after sunset on that day, would show no visible evidence of a snapshot at all. The light level would be diminished by another hundred times, and the brightest stars could be seen. At that point the eye does notice the growing darkness and the contrast between sky and shaded glen becomes large.

At this point we need to say a few words about the measure of light. Two of the most useful terms in light quantification are *luminance* and *illuminance*. In any comparison between the brightness from one light source and that from another, we must be careful to distinguish between the two terms. Luminance is, in the broadest sense, the amount of light propagated outward from a light source; illuminance is, in contrast, the amount of light incident upon a surface.

We can further say that, in general, luminance is equal to the illuminance times the reflectance, the percentage of light reflected from a surface, but to do so raises a complex issue. Here we wish only to introduce the terms. Both of these terms are measured in a bewildering array of units, some defined in the metric system and some in the English system. We have lumens, nits, stilbs, apostilbs, footlamberts, blondels, and candelas per square foot or meter, to name a few quantifiers for luminance. For illuminance we can choose between foot-candles, lux, phots, and lumens per square foot, among other units. Some of this proliferation can be laid to astronomers, physicists, and lighting engineers, each profession using its own preferred units. If that were not enough, comparable sets of units are defined in both the metric and English systems. For example, the foot-candle and the lux are units of illuminance on a surface at a distance of 1 foot and 1 meter, respectively, from the light source, with 1 foot-candle being equivalent to 10.76 lux.

In the remainder of this chapter, we will define sky brightness in terms of candelas per square meter (cd/m^2) or its alternate, nanolamberts (nL), where $1 cd/m^2$ equals about 315,000 nL.

✳

THE STELLAR MAGNITUDE system is a measure of flux, the time rate of flow of light. Stellar magnitudes may constitute the oldest measuring units in continuous existence, owing their origin to Hipparchus, an outstanding Alexandrian astronomer of the second century before the common era. This gifted astronomer was among the first to create a star catalog, the earliest still preserved. He ranked more than one thousand of the stars he could see by brightness or magnitude. The brightest twenty or so were of the first magnitude, the next fainter stars, about sixty, were dubbed second magnitude, and so on down to the faintest stars, which he called fifth magnitude. But after the telescope was developed in the seventeenth century, astronomers became aware that a few objects were too bright to be considered of the first magnitude. Specifically the very bright stars Arcturus, Vega, Capella, Betelgeuse, Rigel, Procyon, and two in the deep southern sky, Alpha Centauri and Achernar, are all of magnitude 0, and the two brightest of all, Sirius and Canopus, are of −1 magnitude. The Sun and the Moon and many of the planets also appear at negative magnitudes.

The five naked-eye planets vary so much in their distance from Earth that their brightnesses can change widely. Thus Mercury varies between magnitudes 0 and –2, but as it is always near the Sun it is never seen in full nocturnal glory. Mars varies even more widely, appearing from –2 all the way to +2 at times. The others are more constant; Saturn appears from 0 to +1, Jupiter is at –2, and Venus is the brightest of all at –4. Finally the Moon varies between –8 as a crescent, to –10 at either quarter when seen half illuminated, to near –12.5 when right at the full phase, and for the Sun we must soar to magnitude –27!

In the nineteenth century, Herschel and Pogson noticed that five magnitudes stand in the ratio of about one hundred to one, and they made this its exact definition. Thus magnitudes form a logarithmic scale, appropriate for the eye and photograph, which detect and respond to light logarithmically. Thus variations of 1, 2, 4, and 8 appear to differ by the same amount while those of 1, 2, 3, and 4 do not. Magnitudes are measured in a logarithmic scale, not to the base 10 or e, but to the base $10^{0.4}$, a number equal to about 2.512. Furthermore, brightness runs counter to positive numbers; the smaller the number, the brighter the star. The relationship between magnitude and flux for two different light sources of magnitude m and m' is given by

$$m - m' = 2.5 \log (F'/F)$$

Thus a magnitude difference of five amounts to a flux ratio of 100. Remember that the system of stellar magnitudes is backward, and so the smaller or more negative the magnitude, the brighter the light source.

Of all commonly observed light sources, none are so faint as the stars. This is easily illustrated in Table 2.1, which lists the magnitudes for common household incandescent lightbulbs and for a candle. At least one of the major manufacturers of lightbulbs prints the number of lumens along with the wattage on each bulb container.

This can be readily transformed into the magnitudes that the bulbs would appear at a given distance. Table 2.1 indicates that all common bulbs and even a standard candle appear much brighter than the fainter stars visible on a clear night. Even the candle outshines at 1 mile all but

TABLE 2.1

The Power Usage and Light Emitted by GE Incandescent Lightbulbs Compared to Stars and Planets

Watts	Lumens	Lumens/Watt	Magnitude at 1 Mile*
300	6,300	21.0	−4.8
200	3,900	19.5	−4.2
150	2,800	18.6	−3.9
100	1,710	17.1	−3.3
75	1,190	15.9	−2.9
60	865	14.4	−2.6
40	505	12.6	−2.0
25	210	8.4	−1.1
15	110	7.3	−0.4
7½	50	6.6	+0.5
Typical candle			+1.8

Planet or Star	Magnitude
Venus at brightest	−4.6
Jupiter and Mars at brightest	−2.6
Sirius	−1.4
Canopus	−0.7
Vega, Capella, Arcturus, Rigel, Saturn at brightest	0.0
Average first-magnitude star	+1.0
Polaris and Mars at faintest	+2.0

*At 1 kilometer, each bulb and the candle would appear 1.03 magnitudes brighter.

the brightest stars. At a distance of 10 miles, each bulb and the candle would appear five magnitudes fainter. All but the candle, even the 7½-watt night-light, could be seen by the sharp-eyed among us on a clear dark night 10 miles away. This emphasizes the reality that astronomers deal with extremely faint light sources, right down to the limit of the eye or the telescope. Far more than aviators, lighting engineers, or perhaps even ophthalmologists, most astronomers know the behavior of the eye and the camera when observing the faintest light levels at the threshold of visibility.

A dark, moonless, unpolluted sky, one in which the Milky Way streams visibly from one horizon to the other, is much darker than the sky at civil twilight. For this, the Sun must be a full 18° below the true horizon, a time known as astronomical twilight. At this point, the last traces of dusk near the horizon vanish and full night begins. Intermediate between civil and astronomical twilight is nautical twilight, when the Sun is 12° below the true horizon. A modest residual glow is still discernible at this point in the general direction of the setting Sun.

The sky is never totally dark. Four separate origins of light are always present in even the darkest skies. Aloft in the upper levels of our atmosphere is the airglow caused by the ambient excitation of atoms and molecules by solar particles, energized as they collide with each other. The airglow is always present but it waxes and wanes over the 11-year sunspot cycle and at its strongest emits the aurora borealis and australis, the northern and southern lights swirling around our magnetic poles.

In addition, debris from comets and other solar system detritus fills the regions closest to the Sun and is illuminated by it. Occasionally we see it directly as a faint glow along the ecliptic and we call it the zodiacal light. Interstellar gas and dust far beyond the solar system is also dimly illuminated by the stars; this and the integrated light from stars too faint to be detected individually form the last two causes of illumination of the black wastes between the stars we see.

Table 2.2 lists, for five altitudes of the Sun and several conditions of night, the sky brightness in units of candelas per square meter (cd/m^2)and nanolamberts (nL). The conversion from either of the two linear units into the ratio of the magnitude system assumes a stellar magnitude of a square degree of dark night sky to emit as much light as one star of magnitude 4.0. Thus a conversion of 17.78 magnitudes to the magnitude of a square arc second is needed. This makes the magnitude of a very dark sky around 21.8 per square second of arc. The data shown here reconfirm the incredible range of adaptation of the human eye to the normal patterns of night and day.

If we compare the brightness of the dark night sky against total blackness, we find that all the light emitting from the half of the entire sky above

TABLE 2.2

The Sky Brightness at the Zenith
for Different Altitudes of the Sun

Sky Condition	Altitude of Sun	Sky Brightness at Zenith cd/m^2	nL
Sky at Midday	40°	1,500	5×10^8
Sky at Sunset	0°	100	3×10^7
Night in Large City	−18°	3	10^6
Civil Twilight	−6°	0.3	10^5
Night with Full Moon	−18°	0.03	10,000
Night in Suburbs	−18°	0.003	1,000
Nautical Twilight	−12°	0.001	300
Darkest Night	−18°	0.0002	54

the horizon packed together would shine as brightly as a very thin crescent Moon, or about six times as bright as Venus at its brightest. This would not include the several thousand stars individually visible to the very sharpest eyesight. The individual naked-eye stars, considered separately in one hemisphere, would together just outshine Venus at its brightest.

At civil twilight, the sky is still about ten times as bright as it is under the Full Moon, as Table 2.2 reveals. One can read print on a white page in moonlight, but cannot detect the colors of objects bathed in its glow. If the Full Moon is well above the horizon and the night is clear, the zenith shines at about 0.03 cd/m² or 10,000 nL. This is still over one hundred times the light from a clear night without the Moon in the sky, but quite a bit fainter than a moonless night sky seen in the center of a city. From the suburbs, where most man-made light comes from unshielded streetlights, the sky is perhaps a tenth the full moonlit sky. Yet even then, the Milky Way can at best be seen only near the zenith and the sky brightens far more quickly with lowering altitude than under natural conditions alone. The clear sky overhead, between midday and midnight, darkens by a factor of some 10 million or, alternatively, about a hundred thousand under Full Moon conditions.

A glance at Table 2.2 shows that a typical suburban sky registers about the same zenith brightness as does one with only natural light at nautical twilight. But the latter illumination is directional, with the light source being entirely in the direction of the setting (or rising) Sun, whereas the suburban sky will show an increase in brightness in all directions from the zenith, unless a bright source such as a shopping center lies nearby.

The number of stars visible to the naked eye depends on several factors in addition to the darkness and clarity of the sky. It is related to the magnitude of the faintest star visible at or near the zenith. But other factors are involved. The photosensitive cells in the retina of the eye come in two very different varieties, called rods and cones. The rods are related to scotopic vision, the night vision that enables the greatest visual sensitivity, and they have no sensitivity to color. The cones predominate at greater light levels, such as daylight and twilight, and provide photopic vision yielding color sensitivity and rendition. This separation serves two purposes and is the reason that color cannot be detected at low light levels or for any but the brightest stars. Only the cones are found at the center of vision, known as the fovea; their concentration there provides maximum detail. With no rods near the fovea, the faintest stars cannot be seen, and observers are advised to use averted vision when viewing faint light sources. But just a few degrees away from the fovea, the rods predominate and the faint stars can be seen.

The crossover from scotopic to photopic vision, sometimes called mesopic vision, is not precisely known, nor does it have a sharp boundary, but it appears to fall between 0.01 and 0.3 cd/m^2 or about 3,000 and 100,000 nL. The sky seen from the center of a city is usually brighter than the lower limit and observations confirm that the sky above reveals some perceptible color.

It appears that the night sky over a major city reaches photopic levels especially in directions away from the zenith. Hence, the number of stars visible in the sky is not closely tied to the faintest visible star. Nonetheless, we can calculate for extinction coefficients of 0.3, 0.5, and 0.8 the number of stars visible at one time in an unobstructed hemisphere of sky. These are the approximate coefficients for a dark sky in the country away from lights, in the suburbs, and in a city, respectively. The average young person may see stars to a limiting magnitude of 6.0, whereas for an older person, the limit is closer to 5.0.

The total number of stars visible at night is not easy to derive. A young observer can spot about fourteen hundred stars brighter than magnitude 6.0 at one time under ideally dark country conditions. The total shrinks to fewer than five hundred for a typical suburban sky and perhaps one hundred from the center of town. Each of these numbers should be reduced by a factor of just over three for a magnitude limit of 5.0, a common limit for a senior citizen. In actuality, the spread from rural to urban skies is usually much greater, due to the presence of direct glare and obstruction in the sky from buildings and trees. On most city streets, the eyes are bombarded by light from bulbs that are directly visible; their glare can reduce the number of stars to a few dozen or fewer.

Some 20 to 30 years ago, an explosion of light took place across the United States and the world, mostly in the form of streetlights and private security lights that were much brighter than their predecessors. They came into widespread use, in a campaign to overwhelm crime with light along with the millions of billboards and yard lights aglow in rural areas all night, every night. This recent eruption of light across the dark side of the Earth is now visible throughout much of the solar system to anyone with telescopes like ours. Today few young people have ever seen the Milky Way, far fewer than had seen it a generation ago. Today their galaxy is taken instead from *Star Wars* and *Star Trek*. The cyberworld may be there but the sense of wonder at the night sky is not. Few today are in a position to see what a dark sky really means.

Images of our world from space by day and by night may have become commonplace; nevertheless they continue to rank among the twentieth century's most influential pictures. One of the best known of them shows our lovely world by day, a full likeness of the Earth, the pearl of the solar system. We see Africa in the center with its deep green rain forests and brown deserts, surrounded on every side by blue oceans and fleecy white clouds. Surprisingly, evidence of human activity makes little visible impact on the appearance from space of the sunlit side of our world. By day, the Earth is strangely devoid of signs of human habitation, even when seen from only a few hundred miles overhead. I recall a satellite image of a portion of China; in its center directly below lay Shanghai, an industrialized urban sprawl of more than 10 million people. It was barely visible—a grayish patch surrounded by an ashen-tinted green and blue. Once in a while on other satellite pictures a dam or a levee might appear as a telltale straight line, but these instances are rare.

Before the space age began, books mistakenly stated that the Great Wall of China would be visible to the naked eye from the Moon, the only human artifact to be so favored. The Great Wall is indeed long enough; it would stretch across half the face of the Moon if built there. But what about its width? Only as wide as a two-lane highway, it is far too narrow to be resolved from nearby space, much less from the distance of the Moon. Many parts of the American interstate highway system would be more easily seen if length alone were the determining factor, yet they aren't visible either.

But once the Sun sets, communities of all sizes are ablaze with light. Slash-and-burn fires destroying the tropical rain forests of the Amazon River basin, fishing fleets in the Sea of Japan, oil-field fires in Siberia and the Near East—all shine forth together, as do few other dark sunless places in the solar system. Like every other large city in the world New York, with its giant skyscrapers and its Great White Way, appears dominated instead by the orange sodium glare of millions of streetlights. From the Moon, the naked eye could only too easily make out the northeastern corridor reaching from Boston to Washington as a patch of light, and cities of the third world appear almost as bright.

It was not always this way; I recall the New York of forty years ago, when I lived in an apartment building on the Upper West Side in Manhattan just a few blocks from Central Park. On a clear summer night from its roof I would occasionally point out the stars to my fellow tenants. Overhead, Vega, Deneb, and Altair were the brightest and they, along with their constellations of Lyra, Cygnus, and Aquila, dominated the zenith all summer long. But now in the much smaller city of Middletown, Connecticut, with our much brighter night sky, I can't often see the outlines of the constellations. Today the three bright stars are too frequently seen alone since their fainter neighbors are invisible. The figure that they form, known widely as the Summer Triangle, is an asterism of very recent popularity. I know personally that this is so and, from four decades of teaching astronomy, I know that brighter skies blot out their fainter stellar neighbors.

"Brighter is always better," proclaimed utility companies, law-enforcement officials, and manufacturers of light fixtures as streetlights and security lights were promoted to municipalities and the public until astronomers, amateur and professional, began to be aware that they were losing their wondrous resource, the night sky. The effects of night lighting on animals and their habitats is evermore the concern of environmentalists. And finally we come to the bottom line. Light pollution costs; it costs billions of dollars annually in the form of wasted money, energy, and fossil fuel.

3

✳

Human Bonding
During a Total Eclipse

"No man is an island, entire of itself; every man is a piece of the continent, a part of the main; if a clod be washed away by the sea, Europe is the less, as well as if a promontory were, as well as if a manor of thy friends or of thine own were; any man's death diminishes me, because I am involved in mankind; and therefore never send to know for whom the bell tolls; it tolls for thee."

John Donne, Devotions. XVII
Ernest Hemingway, *For Whom the Bell Tolls*

On February 26, 1998, two hours after noon, I saw the Moon pass before the Sun, cutting off its brilliant light. I stood in a small field an hour's drive north of Maracaibo, Venezuela's second-largest city and the center of its oil industry. I had gone to Venezuela to converse and collaborate with my close colleague and coauthor of a book on climate and global warming and, weather permitting, to witness this total solar eclipse. He and I, his family and friends, a bunch of other astronomers, another group of geodesists from Germany and Switzerland, some physicists from India setting up their timing equipment, the governor of the Venezuelan state of Zulia and his entourage, representatives of the national media, and many, many others had assembled in a field to watch this mightiest of predictable phenomena of nature. And many thousands lined other fields strewn like beads along a string through portions of Panama, Colombia, Aruba, and various Windward Islands, as well as Venezuela, where the Moon's shadow was due to touch down.

Hundreds of hopeful viewers wandered into this particular field because it fell along the center line of the path of the Moon's shadow, within which the eclipse appears total. The eclipse would be total from Maracaibo itself, but along the center line the totality lasts for the longest time. Signs along highways throughout the region identified the edge of totality as well as the center line. All the hotels in Maracaibo had been booked months, even years, beforehand. Tens of thousands of viewers had arrived, and cars and tour buses crawled along the roads toward the center line that morning, to return to the city later that day.

By midday the morning clouds had lifted. This eased our concern, since the rapid cooling that takes place just before and during totality can sometimes bring on clouds. We were together under a clear blue sky, forming little groups speaking Spanish, English, German, and other languages. Dozens of us here and hundreds farther along were milling around and waiting for the culmination of the spectacle. Telescopes with light filters were being mounted and aligned, in preparation for photography of the Sun at every step of the eclipse. The Indian physicists and the European geodesists hovered over their equipment, making last-minute adjustments. There were snack bars for the hungry and thirsty, and almost everyone had a camera. At noon, some minutes before first contact, when the Sun's disk began to disappear behind the much closer Moon, the normal hot bright sunlit scene blazed away with a normal noontime light intensity on the ground of 10,000 foot-candles, ten thousand times the light of a candle seen at a distance of 1 foot from the eye.

From the Moon just happening to pass in front of the Sun, a most mundane cause, came one of the single most unforgettable sights in nature, a total eclipse of the Sun. The very phrase conveys the consummation of all celestial sights, the king of them all. For three minutes the Sun disappears, and day at once comes unto night. It is here worth recalling that the Earth eclipses the Sun each and every sunset for the rest of the night, but the frequency and the regularity of that event subsumes it into the routine.

A little after one in the afternoon, the Sun was half covered and appeared as a fat crescent. Looking anywhere but directly at it, I saw nothing unusual. Nothing. The light intensity was down to 5,000 foot-candles, with little to show for it. Human eyes are so adaptable to wide ranges of illumination that we cannot recall an exact light level as a memory even seconds after seeing it. This is why partial solar eclipses aren't much of a show. Only for a few minutes before and after the climax of totality could we barely see that the scene was abnormally darkened. Even when the Sun was 99 percent covered, as it was just outside the path of totality, the solar surface was still bright and the exhibit was not very spectacular. Viewing an eclipse while not quite in the path of totality has been rightly likened to being not quite pregnant.

About 15 minutes before totality, the light was reduced to 1,000 foot-candles and the Sun appeared as a thin crescent, about like the Moon when it is first seen in the evening dusk after its invisible new phase. The sky was still a cerulean blue, but now it was beginning to darken perceptibly. It appeared as a scene viewed through sunglasses. Suddenly out in this still-blue sky a brilliant star appeared to the west—it was Venus, a planet visible in a normal clear daytime sky, but not like this! No starlike point should be so bright in the daytime and none ever has been. No, Venus was testimony to something funny, something not right, taking place around us. Time was indeed out of joint. I pointed to the planet, and at once I had everyone's attention. The buzz of many tongues around me began to swell and to focus. Mira! Look! Suche! Hands flew up in a salute toward the brilliant planet.

At a deep twilight level of 200 foot-candles, the birds began to roost. And then the excitement level picked up as the visual stimuli assaulted our eyes with ever increasing frequency.

At 100 foot-candles, only seconds before totality, the faded sunlight cast shadow bands that appeared as stripes of light and dark passing across any light-colored surface. And then, the sky to the west looked dark gray, as if a thunderstorm or even a dust storm of biblical proportions were approaching. This was the Moon's shadow. It formed an elliptical dark area some 100 kilometers wide and long, and moved at twice the speed of a jet aircraft.

The last of the slender solar crescent slipped away and night descended. And we were down to 20, 10, even 5 foot-candles of light. Just a starlike morsel of the disk of the Sun stood out like a jewel, as the ethereal corona became visible on the opposite side of the lunar disk, revealing the aptly named diamond ring effect. And then that diamond, too, was gone.

Irregularities in the lunar surface revealed the phenomenon known as Baily's beads, caused by bits of solar disk shining through the irregular terrain of mountains and valleys on the lunar surface. Bright red solar prominences were seen extending outward from the solar surface and were visible here and there around the dark limb of the Moon. These were bursts of rarefied gas suspended well above the solar surface and thus seen extending out beyond the edge of the dark Moon; their crimson color derives from their emission of light only at one point in the red part of the spectrum, indicative of hot glowing hydrogen gas. And there was the corona, the white uneven rays of light from the tenuous extended upper atmosphere of the Sun, surrounding the Moon on all sides.

Two brilliant planets appeared flanking the Moon/Sun, just 3° on either side; to the left was Mercury and to the right Jupiter. With the white corona in the center of this luminous triad, totality was here; less than a ghostly tenth of a foot-candle of light remained, about twice the brightness of the Full Moon. Bright stars could be seen, and a faint orange-colored band extended around the entire horizon as we could still see into the upper atmosphere outside the shadow, where a bit of sunlight remained to illuminate the haze.

The entire sky seemed somehow finite and of a definite size, like a giant sound stage and unlike the depthless empty blue sky of normal daylight. A feeling of a set piece took hold of us, overwhelming us; we had become children together as one, sharing this unearthly event (see Figure 3.1).

Houston, we have a syzygy! All fell into a grand alignment as the three bodies lined up just as foreordained years earlier. No other science possesses the infallible predictability of celestial mechanics, the subfield of astronomy that deals with the motions of objects in our solar system. No other science has for three centuries perfected a discipline allowing

3.1 Mid-eclipse showing the solar corona. The original of this figure is a color photograph 5.5 by 8 inches, taken by Professor Jeanette Stock, Universidad de Zulia, Maracaibo, Venezuela. It was exposed through a 25 cm reflecting telescope of focal ratio f/10. Courtesy of Jeanette Stock the eponymous Halley to precisely predict his comet's return 75 years into the future, a discipline needing only one modification since it was developed by Newton, the generalization called for by Albert Einstein's relativity theories.

In fact, a total eclipse in 1919 transformed Einstein from a reasonably prominent physicist into a world celebrity of enduring fame. Stars surrounding the eclipsed Sun were photographed at the time, and again months later at night. A comparison showed that with the Sun in their midst, the positions of the stars were distended by just the amount he predicted, proving that gravity does indeed bend light.

Could other sciences someday reach this same level of accuracy? Might future El Niños be forecast with such precision or the degree of global warming and the attendant rise in the sea level around the world? And what a response to astrology this dramatic pageant makes. That moribund discipline, encumbered with its doctrinaire practices dating back into the hoary past; will it ever display such precision and reliability in its predictions?

The 3 minutes of totality came to an end, the diamond ring again appeared for a few seconds, and the scene ran backward. But we had been changed. The shared experience of this disparate lot of humans from around the world brought us together as we had not been before. The eclipse reached into our deeper selves; bringing awe and a sense of the supernatural. Now as the Sun reappeared, all of us, scientists and nonscientists alike, felt we had been members of the same team. We took group photographs of ourselves, to further share and bond together in an act of closure before disbanding.

For the next such sight we would not have long to wait. To be sure, totality occurs somewhere on the globe every year or two, but those in Antarctica or the southern Indian Ocean are not easy to reach. However, on August 11, 1999, the Moon's shadow swept across Europe more surely than did the legions of Napoleon or Hitler, and made a tour of its capitals, passing over or near Paris, Stuttgart, Munich, Vienna, and Bucharest before entering the Black Sea and the Middle East. But the lunar shadow will not pass this way again until August 21, 2017, when that dark spot will race right across the United States from coast to coast.

4

✳

Towers on a Rotating Ball

Nothing could be more obvious than that the earth is stable and unmoving, and that we are in the center of the universe. Modern Western science takes its beginning from the denial of this common-sense axiom.

Daniel J. Boorstin, *The Discoverers*

Falling, falling . . . the massive tower seemed to be perpetually toppling over. Above and beyond it the fair-weather cumulus clouds raced by, driven at an unseemly speed by an impatient wind. They moved in lockstep, these cumulus fractus, as they are called by meteorologists, the ones that appear as tattered remnants of the rounded cumulus fluff balls they may once have been. Not only are they part of the sky, but we perceive the skyscraper as part of it, too. And there lies the problem. Our eyes base the nonmoving frame of reference on the clouds aloft and not on the object thrusting into their midst. The reality is that the tower, not the clouds, is fixed and does not move. Or does it?

The Empire State Building is 1,250 feet tall, not including the television tower on the top of the structure. More than any other tall building built before or since its completion in 1931, the Empire State Building has captured the imagination of the world. Tallest of the three largest skyscrapers designed in the Art Deco style of architecture, it—along with the Chrysler Building and the RCA (now the GE) Building in Rockefeller

Center—forms a triumvirate, not a mile across, of the loveliest elements of the skyline of New York. In the last twenty years it has been exceeded in height by office buildings in New York,* Chicago, Shanghai, and Kuala Lumpur, Malaysia. But none has replaced it in people's fancy as the culmination of the skyscraper (see Figure 4.1).

The building has had its historical moments. King Kong climbed it to try to escape the modern world in a 1933 classic film, and later the great ape fought off biplanes buzzing around him and shooting at him while he held Fay Wray in the palm of his hand.

In July 1945, a young colonel in the army air corps with an arrogant Custerlike attitude and ego to match rammed a bomber into the seventy-ninth floor of the tower, after defying orders to land at La Guardia Airport nearby. After just missing the tops of the RCA Building (seventy stories) and the Salmon Tower at 500 Fifth Avenue (fifty-two stories) in the fog, he flew right down Fifth Avenue and did not miss the tallest tower farther south. Although one of the plane's engines tunneled through the structure and fell out of its far side onto a rooftop, the structure itself suffered only superficial harm. Some attributed its survival to its having been built very sturdily, since its top "hat" was intended to be used as a mooring mast for dirigibles (but never was). At 360,000 tons, it outweighs most newer skyscrapers by a wide margin (e.g., the taller Sears Tower in Chicago weighs but 220,000 tons), although the great pyramid in Egypt, being almost solid, has the record for a structure at 6 million tons.

The elevator shafts in the main tower access all floors up to the eightieth. From that point those who seek its main observatory must change to others that extend from 80 to that perch on the 86th floor. A lone elevator then takes the tourist up the mooring mast to the 102nd floor at the summit.

Once, while making this ascent, I chanced to peer behind an unlocked door to see a stairwell extending down, seemingly forever or at least to the center of the Earth. That gave me pause. I had never looked down a

---

*Author's note: Shortly after this was written, on September 11, 2001, the two tallest buildings in New York, forming part of the World Trade Center, were destroyed and thousands of lives lost by the impacts of two hijacked airliners in the most malicious and unconscionable terrorist attack in history. It is widely assumed that the World Trade Center will be rebuilt.

4.1 The Empire
State Building

longer shaft in my life. I was seized by the youthful urge to spit or drop
something and watch it swirl into the infinite below. I didn't do it, but
later, whenever I consider the Coriolis force, I recall the allure. Not strictly
a force, its alternate designation as the Coriolis effect is more on the mark.
This is the effect that causes wind and missiles to deviate to the right in the
Northern Hemisphere and to the left in the Southern Hemisphere.

The Earth is a rotating near-sphere; hence an artillery shell lobbed
from a gun has its own individual muzzle velocity plus its velocity due to
the gun sharing the rotation of the Earth underneath. The hourly velocity of that rotation at the equator is just equal to the planet's circumference of 25,000 miles divided by 24, or about 1,040 miles per hour
directly eastward. At either pole one would not move, but instead spin

around on one's own axis like a top. At latitudes in between these two extremes, the rotational velocity is 1,040 miles per hour times the cosine of the latitude. At a latitude of 60°, the velocity is half that at the equator, 520 miles per hour. Cities such as Saint Petersburg, Helsinki, and Stockholm lie near this latitude, and westward flights connecting them are just able to match this speed and counteract the Earth's rotation, thus staying in place, keeping the Sun and the stars fixed in the sky. The approximate velocity at different latitudes is shown in Figure 4.2.

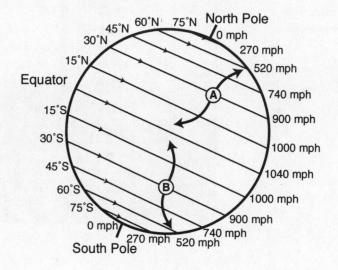

4.2   Latitude vs. rotational velocity

An experienced artilleryman aboard a warship knows to aim a missile to the left of the target in the Northern Hemisphere, allowing for its curve to the right to bring it to the target in any direction from the ship. Wind, too, is subject to the rotation, and the Coriolis effect deviates it to the right just as it does the missile. From this and other considerations, we see wind circulating in a clockwise motion around a high pressure area, a "high," and counterclockwise about a "low." In the Southern

Hemisphere, the reverse is true. It is the rotation of the Earth that gives us these vortices and eddies that circulate throughout our atmosphere.

This fact was put to the test with the early space probes of Venus. That perpetually cloudy world is only slightly smaller than our Earth, and is also now known to be the slowest rotating body in the solar system, taking about 243 days to spin around on its axis just once. Its equatorial rotational velocity is thus only about 4 miles per hour, about the speed of a brisk walk, a power walk with no window shopping. If one could keep up such an effort, one could counter that slow spin, and remain in one spot indefinitely. Of course with an atmosphere mostly of carbon dioxide and ninety times the air pressure at our surface, and a greenhouse effect that maintains the air temperature in excess of 700°F, it is safe to assume that no one will walk on Venus anytime soon.

What does all this have to do with the Empire State Building? It is by no means loaded with carbon dioxide, nor is it excessively hot. Only this; its top, being farther from the Earth's axis of rotation than the base, must rotate about it at a greater speed. At New York's latitude of nearly 41°N, the apex, at 1,250 feet above the ground, is 1,250 times the cosine of 41°, or 943 feet farther from the axis. Each day, the top travels 5,920 feet farther than the base. That amounts to a faster speed of 247 feet per hour, or 0.82 inches per second. The eightieth floor at the top of the long elevator shafts comes to about 80 percent of this total height. Were I to have dropped a stone down the shaft, it would have taken about 8 seconds to reach the bottom, if air resistance were negligible, as it would be for a small dense object. Starting off with a faster motion than the base, the stone would have deviated some 6 inches away toward the east wall of the shaft. With its greater eastward velocity, it would overtake the bottom of the shaft, and thus eventually strike its east wall if the shaft were longer. Since the shaft is wider than 6 inches, the stone would strike a point at the bottom 6 inches east of the center.

The Coriolis effect is that which gives our Earth its appearance from space with an atmosphere of vortices and whirlpools abounding over the entire surface, its intensity varying with wind velocity and latitude. By this rule, Venus should possess almost no Coriolis force and should show no whirlpools in its atmosphere. Indeed that is just what the early space probes confirmed years ago. Below (see Figure 4.3) are satellite images of the two planets to scale. Note the even flow across the face of Venus, especially in direct comparison with our own turbulent world. The flow on Venus is called laminar flow and appears the same as the smoke drifting upward from a cigarette in a quiet room. The smoke flows upward evenly for several inches before breaking out into turbulent eddies. The circulation around Venus is an example of a Hadley cell, named after an eighteenth-century English meteorologist. Our globe was thought to possess this simple circulation model until the effect of its rotation was understood.

At the opposite end we have Jupiter. At eleven times the diameter of the Earth and with a period of rotation of just 10 hours, it has almost two and a half times our angular rotation speed, and its huge thick atmosphere shows a near chaotic swirl of spots and storms all over its disk. The Earth, too, was no slouch a billion or so years ago, when its day was only 8 hours long. The Coriolis effect, and the wind it raised in that distant past,

4.3   The Earth and Venus to scale. Courtesy of NASA

was nearly three times its present intensity with highs and lows careening across the surface of the globe in an unending stream.

And so, in a sense, the tower is falling, falling away from the vertical line of the moment as oriented not with the Earth but with the sky.

＊

THE EMPIRE STATE Building may have escaped experiments of many kinds, but another tower did not. Galileo is alleged to have dropped objects of different weights from the top of the leaning tower forming the campanile to the cathedral at Pisa. If the event happened at all, it probably took place early in the seventeenth century, nearly 400 years after the tower was built. In any case, Galileo was well aware, contrariwise to Aristotle's teachings, that two different weights, both sufficiently dense, would fall at the same rate and strike the ground together. The Greek scientist presumed that the times of falls were proportional to their weights.

Today the Leaning Tower of Pisa tilts toward the south about 5° or some 15 feet away from a vertical line extending upward from the center of its base. This tilt is increasing despite many attempts to stop it. The structure began to lean at the outset, well before it was finished, and the builders compensated for it. Were it now to be righted, its upper floors would lean slightly in the opposite direction. When might it fall over? It would normally be expected to fall at the moment that its center of mass extends beyond the edge of its base. But it is not so rigid that we can wait for that moment. Engineers are well aware of its fragility and know that it would crumble long beforehand. Further efforts to secure it in place are intended to freeze it at its present inclination in order to prevent just that event.

Yet one more famous skyscraper illustrates a recent development in science. This is the Eiffel Tower, which soars 300 meters, 984 feet, above the Champ de Mars and the streets of Paris. Built in 1889 to celebrate the centennial of the French Revolution, it was intended to be torn down not long afterward, as many thought it to be an eyesore. By the end of the Second World War when it survived Hitler's plan to scrap it, it was seen as a national treasure. Gustave Eiffel, who also designed the pedestal for the Statue of Liberty, made of the tower a structure of such slender and attenuated form that it weighs less than the air inside a cylinder just large enough to enclose it. This is a striking confirmation of the fact that air has weight, and a lot of air has a lot of weight. The tower weighs in, according to some estimates, at about 6,500 tons. The weight of air inside a cylinder 300 meters in height and of such a diameter that the tower's square base, 100 meters on a side, would just fit within, scales just over 6,700 tons, more than the structure itself. It is small wonder that its diaphanous appearance has so attracted tourists from around the world.

Air is compressible—very much so. A simple bicycle pump illustrates that. And just as a pump compresses the air under force, so does the atmosphere under the force of its own weight compress in such a fashion that the density falls with height above sea level in a prescribed way. At sea level, a cubic meter of air weighs 1.293 kilograms or almost 3 pounds; 1 cubic foot of air weighs in at 1.3 ounces. Water thus weighs 773 times as much as air at sea level. Still the fact that air is not infinitely compressible becomes a useful feature of its properties. Elevator shafts, now built to just fit the elevator, do not permit a free fall if the car itself should tear loose from its supporting cables and fall. In July 1945, when that bomber struck the Empire State Building, an elevator with its cables sliced fell at such a gradual rate that its operator survived a drop of more than 900 feet.

The lightness of the Eiffel Tower results from its remarkable tenuous construction. The cross sections of its network of girders and beams of ever finer detail resemble nothing so much as a fractal structure. *Fractal* is a word coined by Benoit Mandelbrot, a French mathematician now living in the United States. Fractal patterns repeat their basic shapes and properties at any scale; imagine the coastline of a continent mapped onto a single chart. The coastline is irregular, having projections of land and indentations of water of no regular shape or pattern. Now shrink the scale such that one small coastal country fills the page. Its shoreline shows just as much irregularity as that of the larger map. Another reduction to perhaps a single bay or harbor continues in this manner, as will any number of further reductions. In the case of the cross section of a tower, the reductions might resemble the illustration seen in Figure 4.4. The second diagram shows a beam with only $\frac{5}{9}$ as much material as the first, yet its strength and rigidity will hardly be diminished. In the third diagram a repeat of the pattern occurs and now only $(\frac{5}{9})^2$ or $\frac{25}{81}$ as much material, less than a third of the original, remains. In this manner the beam becomes much lighter but with an almost undiminished tensile strength. Gustave Eiffel didn't realize then that he was working along the same lines as Benoit Mandelbrot and other contemporary mathematicians a century later.

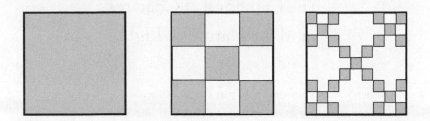

4.4   Fractal patterns

Towers reveal further points of interest. No large towers resemble each other as much as do the two towers of a suspension bridge. But they cannot be erected parallel to each other. Each must be aligned so that it points directly upward, away from the center of the world. At 751 feet in height and separated by 4,200 feet, the two towers of the Golden Gate Bridge connecting San Francisco with Marin County to the north are positioned such that their tops are farther apart than are their bases, due to the sphericity of the Earth. For each to be truly vertical, the tops should be about 1.8 inches farther apart.

The two longest spans in the United States, the 4,200-foot Golden Gate Bridge and the 4,260-foot Verrazano or New York Harbor Bridge, are now dwarfed by the newly completed Akashi Kaikyo Bridge near Kobe, Japan. With a span of 6,500 feet and tower heights of 984 feet (exactly the height of the Eiffel Tower), this behemoth requires a differential tower separation of about twice that of the Golden Gate, or about 3.67 inches. The eye cannot detect such a small increment as this, but it is a critical factor in the design of suspension bridges. Few objects combine beauty and utility as do suspension bridges, but they require extreme care in their design.

Towers and bridges share one other characteristic—since we usually look up at them, they form part of our sky.

*

5

The Cathedral at Chartres
and the Nature of Light

When you can measure what you are speaking about and express it
in numbers, you know something about it; but when you cannot
measure it, when you cannot express it in numbers, your knowledge
is of a meagre and unsatisfactory kind: it may be the beginning of
knowledge but you have scarcely in your thoughts advanced the
stage of science.

William Thomson, Lord Kelvin

Light is funny stuff; either it is made up of a stream of particles called
photons or it comprises a series of waves, or it is just possibly both. Scientists can't quite decide which of these conflicting scenarios is the correct one, but they agree on its speed.

In a vacuum such as empty space, light moves 299,792,458 meters
per second, precisely and by definition. That's about 186,282.396 miles
per second, and nothing goes faster at any time ever. Since we now know
(and hence define) the value of the astronomical unit (the mean distance
between the Earth and the Sun) to be 149,597,870.66 kilometers or
92,955,807 miles to the nearest mile, we also are aware that light takes
499.0047815 seconds to reach us from the Sun, or about 8 minutes and
19 seconds. I use these very precise numbers advisedly, to reveal the
degree of precision upon which modern physics is based. The meter is,
exactly and by definition, one part in 299,792,458 of a second of time.

That means that in that part of a second, the distance light travels in a vacuum is a meter and defines the meter. Since the inch is exactly and by definition 2.54 centimeters, the American system of feet, miles, and pounds is also defined by the speed of light. Albert Einstein realized that $E = mc^2$, where E stands for energy, m is mass, and c is the speed of light, m and E being different manifestations of the same thing. In this most famous of all equations, matter, energy, and the speed of light are all intertwined in a fundamental way. Thus c is of overwhelming importance throughout the universe.

Here we have time and the most important of all distances, the astronomical unit, defined in terms of each other and interwoven together with extremely high accuracy, such high accuracy that they now define much of the rest of physics. These results brought about a change in the definition of the meter, the basic unit of length. No longer is it defined by the distance between two marks on a certain meter stick in Paris. Since 1972, it has been defined in terms of the speed of light and the light time to the Sun.

Since the Earth is known to be an erratic timekeeper, it no longer serves to define time. Its rotation period varies due in part to a secular slowdown of its rotation by about 1 millisecond per century. The tides raised by the Moon and the Sun act as a brake steadily bringing this about. There are also periodic seasonal variations that every 1 year and 6 months slow down and speed up the rate. Then on top of this, we find

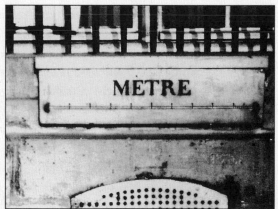

5.1 Meter stick in Paris, formerly defining the length of 1 meter

irregular variations caused most likely by motions deep inside the Earth itself. Thus for very precise work, the Earth simply will not do, and scientists now define 1 second as amounting to 9,192,631,770 periods of resonance of a specific radiation of an atom of the element cesium 133.

Light moves fast, so fast that in our everyday experience we can think of it as instantaneous, and it is—almost. If you are walking in the moonlight along a sidewalk constructed typically of concrete squares 6 feet on a side, and if a ray of light were to leave the Moon at the exact instant you pass over a crack between one block and the next, you and that ray will meet at the following crack 6 feet farther along the sidewalk, both taking about 1⅓ seconds to get there.

The other obvious property of light is its variation in color. Colors differ only in wavelength, a single parameter, as shown by the curve called a sine wave. One wavelength is the distance between one crest and the next (or one trough and the next) and is symbolized by the Greek letter lambda, or λ.

Our eyes see the change in λ as a change in color. The colors we see are the colors in the spectrum that appear whenever sunlight passes through a prism. As Sir Isaac Newton first discovered, white light is broken up into colors, because each color or λ refracts through it at a slightly different angle. The spectrum or rainbow is continuous without breaks, but we customarily denote them as red, orange, yellow, green, blue, and violet, sometimes collectively abbreviated as ROYGBV. Occasionally indigo is considered to be a separate color between blue and violet. In this case the abbreviation is ROYGBIV, but in either case this is the order from longest to shortest wavelength with red light being almost twice the wavelength of violet light. These wavelengths are very tiny, red and violet light having wavelengths of only 0.0007 and 0.0004 millimeters, respectively. In order to avoid such tiny decimals, scientists customarily express wavelength in nanometers, where 1 nanometer (abbreviated nm) equals a billionth of a meter and thus a millionth of a millimeter. Red and violet light turn out to measure near 700nm and 400 nm, respectively, and yellow light, where our eyes are most sensitive, falls near 550 nm, right in the middle.

Light between these two limits in wavelength is visible light, to which our eyes are sensitive. But beyond the red end of the spectrum light of even

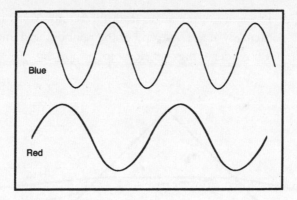

5.2   Relative wavelengths at blue and red colors

longer wavelengths exists, to which our eyesight is insensitive. These range from the infrared out to radio waves that include the AM and FM radio frequencies among others. Wavelengths shorter than the violet range through the ultraviolet down to the lethal X rays. Physicists refer to the whole business as the electromagnetic spectrum, in order to differentiate it from the small section of visible light. Although the window of visible light accounts for only a small fraction of all possible wavelengths, more than 92 percent of all sunlight falls in that window, and it is probable that eyesight evolved to just those colors. All of the hues we see—chartreuse, mauve, russet, hot pink, robin's-egg blue, and all the shades of brown—are just blends of light from different places in the visible spectrum. Think of the big jumbo set of crayons you had as a child—each and every color comes from that spectrum. And as you may recall from those days, a mix of all colors from crayons or paint always turns out to be a kind of sickly greenish brown. Nonetheless ROYGBV gives rise to all of it.

The sky is full of hues and tints of all kinds, especially when clouds are present. Most particles of air are of just the right size to scatter the shorter, the blue and violet, wavelengths, while letting the red and orange light pass through largely unimpeded. Thus the clear daytime sky appears blue to us. As you let your gaze wander from the zenith, the point directly overhead, down toward the horizon, you are looking through ever more air. The air mass through which light from the Sun, Moon, and stars passes is just a measure of the angle from the zenith. More air scatters more light, particularly at the blue end of the rainbow, and thus the Sun and Moon appear redder as they get closer to the horizon.

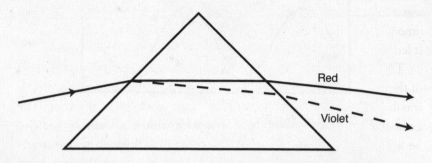

5.3 Prism showing refraction at extremes of the visible spectrum

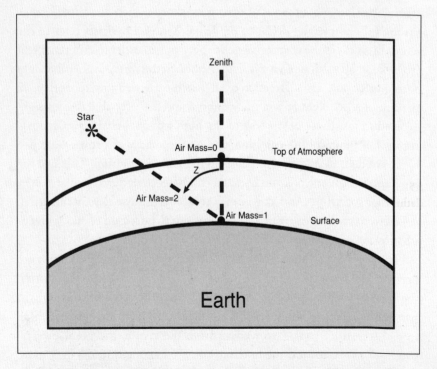

5.4 The dimming of starlight depends on the air mass and the air quality.

All of us recall particular scenes in which the lambency or the colors were unusual or even spectacular. One of my favorites took place one sunny summer day in 1981 in the French city of Chartres. There, at midday under a sky filled with fleecy cumulus clouds, my father and I made our way from the railway station to the magnificent Chartres Cathedral. This was a day of fair-weather cumulus clouds passing in front of the Sun from time to time, only to let the full sunlight pass between them moments later. It led to a transcendent moment both inside and outside this structure.

This grand edifice lives up to its reputation as a heroic work, largely of the twelfth century and of finer construction than later cathedrals. It was built of the hardest and heaviest stone available at the time. The cathedral and its two dissimilar spires rest astride a hill, said by some to be a place of worship of Druids and other pre-Christian religious services. The south tower, the older, stands 106.5 meters or 350 feet above the church floor, and the north tower is hardly less tall.

5.5   The Chartres Cathedral

We entered the church and the light level darkened accordingly. Our eyes needed several minutes to accustom themselves to the much darker interior. Developed in northern Europe, the Gothic style replaced the earlier Roman style, with its smaller windows: the huge Gothic windows are a crowning feature of the cathedral and provide more light, especially on the brief dark gray days of the long winters. The spectacle of that crepuscular interior was great enough under any conditions, but as we experienced the light diminish and brighten again as clouds passed in front of the Sun, the scene took on an unearthly beauty.

The great structure reminds one of a different culture, one honoring the Blessed Virgin, which Henry Adams, in his autobiographical account of his education, adopted as the symbol of the twelfth century, a symbol not only of reverence but of power, a motivating power that induced men to erect such wonders as this. His contrast with the dynamo as the counterpart totem of his own nineteenth century only tends to ratify the Virgin's position in the earlier age.

*

ALMOST 50 YEARS ago, a book was published by Marcel Minnaert, a professor of physics at the University of Utrecht in the Netherlands, which deservedly remains in print. It bears the title *The Nature of Light and Color in the Open Air.* Professor Minnaert, obviously an ardent lover of nature, takes the reader on a tour of many features that appear in the sky, including rainbows, mirages, halos, coronas, iridescent clouds, and many other phenomena that sometimes are seen in or near the sky. There is much here for artists and photographers as well as scientists. It is beyond the scope of this book to cover more than a few of the sky's many treasures, but we will touch upon some of the most noteworthy among them.

The water of lakes or the sea reflects sunlight or moonlight (or even light from Venus as is mentioned in Chapter 1) in a variety of ways, depending on the wind and waves and the angular height of the light source above the horizon. The impressionist painters modeled the effect of wave action upon the waters, and J. M. W. Turner is noted for his faithful reproduction of the columns of reflected sunlight on the Thames

River in smoky London. In the main, the breadth of the solar reflection in the waters is a function of the altitude of the Sun above it. The higher the Sun in the sky, the broader the region of reflected sunlight, although the roughness of the water also affects its width.

The refraction of light by our atmosphere gives rise to many observed phenomena. Light from a star seen at the zenith is not bent in any direction but comes straight down. But away from the zenith, the light is bent always in such a way that a star is displaced toward the zenith and away from the horizon. The amount of refraction is very small for altitudes well above the horizon and depends on the color or wavelength as well as on the altitude. In simple terms, the refractive index, R, corresponding to the angle of uplift of the image, is equal to the difference between the true and the apparent angles from the zenith. Thus

$$R = z(t) - z(a)$$

where $z(t)$ and $z(a)$ are the true and the apparent zenith angles. At an angle of 45°, yellow light ($\lambda = 550$ nm) is refracted upward by just about 1 minute of arc, $\frac{1}{60}°$. (For convenience, the Sun and Moon subtend angles of about 30 minutes, or $\frac{1}{2}°$.) This is inconspicuous to the eye, but near the horizon, the angle is far greater. Right at the true horizon it is about 0.6°, yet just half a degree higher it is reduced to 0.5°.

When the Sun (or the Moon) is just above the true horizon so that the lower limb or edge just touches the horizon, the whole Sun is actually *below* it. The lower limb is lifted 0.1° more than the upper limb, so that the Sun appears squashed into an oblate shape. Since the image appears higher, the effect is to increase both sunrise and sunset by about 2 minutes. The day is actually some 4 minutes longer than it would be if we had no atmosphere, as is the case on the Moon.

The atmosphere also diminishes incoming light from stars or anything else above it, sometimes by a fairly well known amount. On any good clear moonless night away from city lights and haze, the sky appears to become less crowded with stars near the horizon because one is peering through much more atmosphere, as we mentioned above. A glance at Figure 5.6 shows how this works. If we define the amount of air between the observer and his zenith as one air mass, then the amount between him and a star in any other direction is the secant of the angle between the star and the zenith.

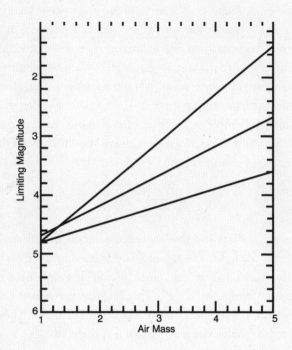

5.6 Extinction as a function of secant z, where z is the zenith angle. The straight lines show, from top to bottom, typical extinctions of 0.8 magnitudes (the case for city centers), 0.5 (the case for suburban areas), and 0.3 (rural areas).

The diminution of the brightness of a star is set by the air mass and a factor called the extinction coefficient, which depends on a number of features of the air. These include haze, smog, smoke, light pollution, and elevation of the observer above sea level. At a sea level sight on a clear moonless night, the extinction is about 0.3 magnitudes at an air mass of 1 (at the zenith). We assume horizontal homogeneity through the air; clearly if clouds or haze lies in one direction, the stars will be abnormally faint in that part of the sky. Air mass 2 occurs whenever an object lies 60° from the zenith; that is, 30° in elevation above the horizon. Air masses of 3, 4, and 5 are found at altitudes of about 20°, 15°, and 12° above the horizon. A star is fainter by 0.3 magnitudes at 2 and that much fainter for each extra air mass. Below these elevations, the rule doesn't quite hold, because the Earth is round. This also tells us how bright the stars would appear in nearby space above the atmosphere or on the night side of the Moon, for out there, the air mass is quite clearly zero. We know that stars appear 0.3 magnitudes brighter there than from sea level on the best of nights.

At the elevations of the large mountaintop observatories in Arizona and California, the extinction is about 0.2 because from their sites of roughly 6,000 to 8,000 feet above sea level, about one third of the atmosphere lies beneath the summit; the air pressure is about two thirds that at sea level. And at the telescopes on the 13,800-foot summit of Mauna Kea in Hawaii, the extinction might be near 0.1 magnitude.

Air is never totally transparent but if more aerosols—more particulate matter, man-made or natural—are present, the extinction increases. Not only does dirtier air block more starlight, but light pollution shining from below illuminates these aerosols and the brighter sky increases the glare through which stars must be seen. On the best nights in a typical suburban neighborhood the extinction might be about 0.5 magnitude, unless the streetlights are shielded from above and the neighbors keep their private lights under control. In the city closer to the center it might be 0.8, 1.0, or even greater.

All of these figures are meaningless if direct glare is present—if one or more bright bulbs shine directly into the eye. Just one streetlight or porch light can blot out many of the stars that might otherwise be seen.

And just how many stars can be seen on a particular night? This is not easy to determine for many reasons, as noted earlier. Eyesight varies from one individual to another, with one's age playing a large role. My own experience is typical. At about thirty, on the best of nights, I could just see the planet Uranus. At magnitude 5.7, it is just above the 6.0 nominal limiting visible magnitude given in most sources. Now over sixty, I can just see 61 Cygni, a well-known double star whose combined magnitude is 4.8. Successful cataract implant surgery has restored my eyesight; my eyesight is about average for my age.

This typical shrinkage of about a magnitude has a tremendous effect on the number of stars visible. With each whole magnitude fainter, the number of visible stars triples! For the number of stars brighter than 6.0 over the entire celestial sphere the common figure is six thousand, with twenty-five hundred visible at one time (not three thousand because of the shortage of stars near the horizon). However, these data need revision downward, as indicated in Table 5.1 below. Here the number of stars in the entire sky is given along with the average number visible on a dark night from one site. The numbers are not of high precision because the numbers vary considerably due to seasonal variation and with latitude on the Earth. The central regions of the Milky Way are quite far down in the southern sky; hence, the stars visible from our northerly latitudes are fewer than are visible from south of the equator.

The numbers are revealing; I could spot some one thousand stars at a younger age and only some three hundred to four hundred today. It is worth noting that when Hipparchus, the eminent Alexandrian astronomer, compiled his catalog of all visible stars, he included just over one thousand, a total consistent with these data. Unfortunately, we do not know his age when he did this.

TABLE 5.1

Number of Stars Visible Brighter Than a Given Magnitude

Magnitude Limit	Number of Stars in the Entire Sky	Average Number of Stars Visible from the Earth's Surface at One Time
3.0	146	37
3.5	278	66
4.0	530	120
4.5	930	215
5.0	1,640	400
5.5	2,820	730
6.0	4,850	1,320
6.5	8,330	2,370
7.0	14,300	4,200

*

How Far Is Up?

The two most widely separated things in the universe are "should be" and "is."

Anonymous

How far is up? How big is our universe? People have posed these questions since the dawn of curiosity. But the first serious attempt to provide an answer was made at Alexandria, the great Egyptian seaport and center of learning, in the third century before the common era by Aristarchus, among others. It was then and there, as much as at any time and place, that the need for quantification was born.

The distance between the Earth and the Sun is perhaps the best-known large number in astronomy, as well it should be. This is approximately 93 million miles or 150 million kilometers and is the yardstick for the distance to any object in the universe except the Moon. If we have it wrong by even a few percent (and we don't), everything from the speed of light to Edwin Hubble's expansion rate for all those distant galaxies goes into the waste basket and astrophysics would have to start over. As one may imagine, an enormous effort by many scientists has gone into as precise a measure of this distance as possible. Since our orbit is basically elliptical, the distance to the Sun varies in the course of the year by about 1.5 percent on either side of the average, the number we commemorate. For convenience and to avoid overuse of large numbers, we define this average distance as 1 astronomical unit (or 1 AU).

Since about 500 B.C. people such as Pythagoras, Plato, and Aristotle knew that the Earth is round and were well aware of the nature of the Moon's phases and eclipses. But it was Aristarchus (ca. 310–230 B.C.), an Alexandrian astronomer, who began (as much as any one person did) the measurement of planetary distances and sizes. He used triangulation rather elegantly to find that the Moon was about 60 times as far from the Earth as its own radius. He knew that anything that subtends about 0.5° in angular measure in the sky, as is the case with both the Moon and the Sun, must be around 115 times as far away from the observer as it is in diameter. Therefore he knew the size as well as the distance to the Moon in Earth terms.

His ablest successor, Eratosthenes (ca. 276–195 B.C.), managed to measure the circumference of the Earth pretty accurately, probably to within 10 percent and possibly to 1 percent of its modern value. That gave the Alexandrian scientists the approximate correct sizes of the Earth and the Moon and the distance between them, in units such as miles or kilometers. But the determination of the size and distance to anything beyond the Moon was a different story. No one managed to get that right until only about 3 centuries ago, not surprising in view of the fact that no planet comes within one hundred times the Moon's distance at any time.

Those Hellenistic astronomers tried in some very imaginative ways to pin down the distance to the Sun. Aristarchus reasoned that at the precise moment of first- or last-quarter phase, the terminator—the line between day and darkness on the Moon—is a straight line, thus making an exact half-moon of light. He was also aware that one angle of the three in the triangle formed by the Moon, the Earth, and the Sun at that moment is a right angle of 90° in size. But this right angle lies not at the Earth, but at the Moon (see Figure 6.1). The other two angles together add up to another 90° since all three angles must total 180° in a right triangle. This is a sound method and Aristarchus tried to measure the angular distance between the Sun and Moon whenever the terminator defined a half-moon.

This is not easy to do. It must be done in the daytime when the Sun is still above the horizon. Glare from the Sun strains the eye and other pitfalls also exist. But Aristarchus tried his best and obtained a value of 87° between the Moon and Sun, and from that the Sun turned out to be nineteen times as far from us as the Moon. This made the Sun almost five times the diameter of the Earth—and the event marked the first time another celestial object was believed to surpass our world in size.

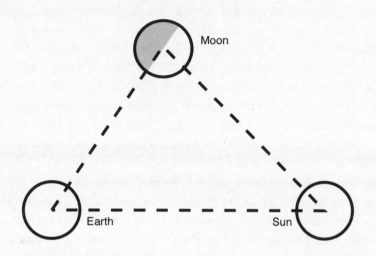

6.1   The distance of the Sun as determined by Aristarchus, who found an angle of 87° at the Earth and a 90° angle at the Moon

The Alexandrian astronomers went further; they derived a size and distance for each of the five bright naked-eye planets that shine in our skies. These are Mercury, Venus, Mars, Jupiter, and Saturn. With the Sun and the Moon, they gave the number seven its special significance and gave names to the seven days of the week. The Alexandrians of course did not understand that the real center of the solar system was close to the Sun (although Aristarchus suggested just this, as Copernicus was well aware when he put forth his own theory 18 centuries later), and they had a garbled order for the planets as a result. But they correctly placed Jupiter beyond the others and Saturn at the edge beyond Jupiter. The stars were thought to be plastered on some kind of invisible sphere just beyond Saturn, then the farthest known planet.

By triangulation among other means, they built a logical cosmos, whose grand form was thought to be a sphere just about the correct size of the orbit of the Earth, or twice 93 million miles in diameter. No one seems to have thought much about what was beyond that limiting distance. A round finite plenary universe filled with the quintessence, the pure crystalline unearthly substance of Aristotle, leaving no vacuums, and centered on our imperfect globe made of earth, air, fire, and water in some mix, fulfilled their need to know. Subsequent observations, made over the next 10 centuries by the Arabs in the Middle East and others, called for a bit of fine-tuning among the planetary distances, but the great system fashioned in its final complex form about A.D. 150 by Ptolemy, the last of the ancient Greek astronomers, held sway for more than a millennium. A sphere some 200 million miles in diameter held all there was, is, and ever will be, with God or gods responsible for it and for keeping it in motion. That a specific size was devised and later foreordained was a notable concept when compared to the vague cosmologies that preceded it.

It is a mistake to disparage the final system devised by Ptolemy because it was centered on the Earth. It was an incredibly complex system of mathematical finesse and accounted for planetary motions with great precision. I am reminded of the final movement in Mozart's last symphony, Symphony no. 41, or the *Jupiter* Symphony. It was the last symphonic movement he was to write and is a marvel of contrapuntal intricacy. Mozart introduces four themes, one after another, but in the development they recur, simultaneously—first two, then three, and finally all four together in perfect harmony. With all the difficulties inherent in a metaphoric comparison of a scientific theory to a symphony, I am still impressed at the suggestive similarity of details blending into a comprehensive whole.

Then after a millennium something funny happened. Along about the thirteenth century European scholars pored over the ancient writings of Aristotle and others that had only at that time been made available in Latin, the lingua franca of those who could read. These sources caused them to think, to ponder again about the kind of universe we have. Babylonian and later Middle Eastern observations of greater consistency and precision than those available to the Greeks also became available. They could be combined with the old theories, to ponder again our world. This led straight to Copernicus who came up with his alternate blueprint, published in 1543. Copernicus was not the first to speculate that the Earth moved. Nicolas of Cusa did that as early as 1440 when he stated, "I have long considered that our planet is not fixed, but moves as do all other stars." And Aristarchus is known to have posed the problem many centuries earlier. But Copernicus laid out a mathematically sophisticated system that accomplished Ptolemy's goals more simply, thus calling into play Occam's razor and its implications. A fourteenth-century philosopher, William of Occam or Ockham, laid down a useful scientific principle that assumptions used to explain a phenomenon should not be multiplied beyond a necessary minimum, that the simpler of two competing theories is to be preferred, all other factors being equal.

Now came the big dilemma—which of the two systems, one centered at the Earth and the other at the Sun, was right? Both could not be correct and one had to go, but which one? Then about 1570, the wealthy Danish astronomer Tycho Brahe offered a third, a compromise with some of the features of Ptolemy and some of Copernicus. Each of the now three systems had its faults and virtues and supporters and detractors. Galileo's strong advocacy of the Copernican arrangement pushed the matter into the forefront of science and religion, changing both of them forever. Galileo had no conclusive proof, but he did make one discovery after another with his telescopes that tore away other portions of the then extant cosmology of Ptolemy. In the extended triumph of heliocentricity over geocentricity, science underwent a pruning from the exhausting discourse on all minutiae as is described, for example, in *The Name of the Rose,* Umberto Eco's story of fourteenth-century monastic life. In it the arguments over whether Jesus ever laughed or not continued on interminably. From Galileo and others, science learned later to progress to a wary acceptance of the theory that best explained the available observations.

The point that (almost literally) blew the universe open was that for the Earth to orbit the Sun in the Copernican theory then steadily gaining favor, the stars had to reflect its motion by moving back and forth every six months. Stars were not seen to do that. From Aristotle to Tycho and Galileo, many sought to observe this feature, known as parallax, all without success. Could it be that the stars are not just beyond, but hugely, exceedingly, so far beyond Saturn that the parallax is there but too small to measure? In that case, they are so distant that they aren't just mere planets as had been thought, but are rather like the Sun, and that makes the Sun, after all, a star.

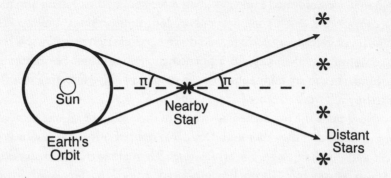

6.2 Trigonometric parallax of a nearby star using the Earth's orbit as a baseline

Galileo's contemporary, Johannes Kepler, formerly one of Tycho's assistants, puzzled over this. He took Tycho's 30 years of observations, the most precise ever made before the telescope, and tried for years to reconcile with them all three of the prevailing systems. None worked. With his gifted mathematical genius, Kepler finally understood the problem. Everyone, even Galileo, believed that orbits in the form of the perfect circle were the only ones possible. But Kepler kept going back to the drawing board, since Mars, above all, could not be made to track along any circle. Finally Kepler thought to try elliptical orbits and ellipses worked. He threw most features of all three systems—Ptolemaic, Copernican, and Tychonic—into the theoretical Dumpster forever. His arrangement with elliptical orbits around the Sun at one focus of each ellipse was the theory that fit the available facts.

The Keplerian universe closely resembles our modern one; we still use his blueprint with only minor reshaping. But here we come back to the original problem: How far is up? Kepler was the first to find the correct relative sizes of the orbits of the planets, but he had no yardstick. His problem can be likened to a map of Europe or America, showing each state or country in its correct shape and place but without the scale; he didn't know how many miles there are to the inch. Neither he nor anyone else knew the size of the astronomical unit.

Kepler died in 1630 and Galileo in 1642, the year Isaac Newton was born. It was Newton who put the whole thing together. Kepler supplied the how in planetary motion and Newton the why. Kepler knew that elliptical motion meant that planets get closer and later farther and still later closer again, to the Sun. This meant that something other than clockwork once started by God or gods sufficed to account for the motions he explained. In 1687, Newton published his *Principia Mathematica,* quite possibly the greatest intellectual effort ever made by a single human being. In it he stated that the force holding the planets at bay was an attractive force of gravitation that weakened with the inverse square of the distance between the centers of the Sun and the planet. To prove this, he needed the calculus, a powerful but then unknown tool of mathematical analysis, so he invented it. A German mathematician by the name of Leibniz made the same discovery independent of Newton. This led to a notorious row in which neither scientist, nor their mutual monarch, George of Hanover and by that time also George I of England, nor any other participant, emerged with honor.

Newton claimed that his laws of motion and gravitation were universal, and observations quickly proved him right. That the same laws apply on Earth and throughout the solar system, and in the systems of stars far beyond, was a philosophical breakthrough of the very highest order. At once, Newton became a worldwide celebrity, and, some contend, the most influential person of the previous millennium. His laws have been modified and generalized just once; in doing this, Albert Einstein also became a worldwide celebrity and *Time* magazine's Person of the Century.

One of the scientific developments that led to Kepler's and Newton's success was the discovery of the correct size of the orbit of the Earth. It was not easy, but from the time of Kepler until that of Newton, the size of the astronomical unit went from Ptolemy's and Copernicus's estimates of about 1,200 Earth radii (e.r.) to something much larger.

Since astronomers still didn't know the exact radius or diameter of our globe, they retained its radius as the benchmark distance. Thus 1,200 e.r. equals around 5 million miles, though not precisely that length. Knowledge of distances to celestial objects remained qualitative and speculative in most cases. But throughout the seventeenth century, the telescope was changed from a thing offering glances of the planets to a device for very accurate measurement. The micrometer was developed, providing crosshairs in the telescopic field against which the diameter of a planet could be measured.

At last astronomers had a means for reliably extending to the other planets Aristarchus's triangulation method for measuring the Moon's distance. Kepler had provided the relative distances to planets for any epoch; now with the micrometer those data could be extended into the domain of the absolute. The method is the same as that for the parallaxes of the stars, using the Earth's orbit as a baseline, but it uses the much smaller diameter of our world, which, with the micrometer, yielded distances to the other planets, if not to the stars. The size and distance to any planet gave us the same information for all the others and for the Sun. Observations of the Sun itself were impossible in the brilliance of daylight, but the nearest planets—Mercury, Venus, and, above all, Mars—were observed at night when their disks could be well measured.

By 1670, various observers had placed the Sun anywhere between 7,300 and 25,000 e.r. The solar parallax—the angular size of the Earth's radius as seen from the Sun—ranged from 30″ (seconds of arc) to 8.2″ (recall that the smaller parallax translates into the larger distance). This meant that the Sun was between 30 and 100 million miles off. Soon afterward, simultaneous observations of Mars made by Giovanni Cassini in Paris and others at Cayenne on the northeast coast of South America, produced 9.5″ for the parallax and correspondingly 87 million miles, a distance too small by less than 7 percent, but close indeed compared to all preceding attempts. Other astronomers of the time found nearly the same result. Thus Newton had a pretty good idea of the scale of the solar system, which Kepler knew better than Copernicus and his antecedents, but not too well. Since Newton's time the uncertainty in the astronomical unit became less and less, so that today this all-important distance is known with an error of only about 1 mile! NASA simply could not have placed the *Voyager 2* space probe between Neptune and its large moon, Triton, nearly 3 billion miles away from us, if the astronomical unit were at all uncertain.

＊

WITH THE CORRECT scale of the solar system in hand, scientists turned to the stars. The Sun-centered model had been fully accepted and the 2 AU baseline of our orbit was used time and again to detect the stellar parallax. It had to be there but no one could find it for even the closest stars until 1838. Then, around 2,200 years after Aristotle sought it, three astronomers measured the first extra-solar-system distance, each independently of the others. The triple discovery clearly shows that it was then that telescopic equipment became sufficiently accurate to observe the distances to the stars. By the First World War, distances to thousands of stars were known. How far was up known to be at that time? Not less than several thousand light-years. A light-year—the distance light travels in a year—amounts to about 6 trillion miles or 10 trillion kilometers; the very nearest star, at 4 light-years, is more than seven thousand times as far as Pluto, the farthest planet. The stars collectively were found to inhabit a disk-shaped volume of space centered on our solar system and extending for thousands of light-years along the plane of our galaxy, the Milky Way, and a lesser distance perpendicular to it.

So things stood until 1920. On April 26 of that year a spirited debate occurred at the National Academy of Sciences in Washington, D.C., between Harlow Shapley, director of the Harvard College Observatory, and Heber Curtis, director of the observatory of the University of Michigan. Studies of the stellar system we call our galaxy led to conflicting conclusions about its size and the location of its center. Shapley contended that the Milky Way was not the prevailing model extending several thousand light-years from the Sun, but was rather more than 100,000 light-years across. Furthermore, he maintained that the center lies about one third that distance away, placing us well out in its suburbs. Observations over the next few years proved him right on both counts.

But Curtis also scored a victory. With ever-larger telescopes, astronomers had come to recognize dozens of spiral-shaped nebulae, faint extended light sources like the more familiar nebulae of irregular shape. Shapley felt that the spiral things were nebulae well within and belonging to the Galaxy but Curtis claimed that they lay well outside it, and indeed formed island universes of their own. Until that time the universe and the Galaxy were held to be identical, different names for the same entity. Now Curtis correctly showed that this was not the case; the universe was much larger and contained much more than just our Milky Way.

So we ended up being shifted away from the center of God's universe once again, this time by almost 2 billion times the shift of Copernicus from the Earth to the Sun. That wasn't our only indignity. Curtis had to go and show that our galaxy is only one of many and no one knows where the center of all that lot might be. These findings made the front page of the *New York Times,* but the debate surrounding them was left to the astronomers. We had moved far from the theology that got tangled up with the debate over the Ptolemaic and Copernican concepts, and science and religion showed each other more mutual respect.

Our perception of the universe continued to grow and has not yet stopped growing. Larger telescopes and auxiliary equipment of greater sensitivity reveal ever fainter and more distant galaxies. Shortly after the debate, Edwin Hubble made his epochal discovery that the velocity of recession of a galaxy is directly proportional to its distance. Now, for the farther among them, we have but to measure this velocity to set the distance. The farthest we know to be some 15 billion light-years away, and if the current big bang theory is correct, the universe cannot be much larger than that. We must be aware in all this that we are not going to come to a wall or boundary of any kind whatsoever, just as on the Earth where one could walk or swim (ideally) forever and never come to a wall. There is only so much Earth; we say that it is finite but unbounded. So, too, the universe is likely to be finite but unbounded. We just can't picture the extra dimension involved. After all, a truly infinite universe is probably even harder to perceive, even though it means that it is certain there is and must be a baseball team up there somewhere that can beat the New York Yankees!

So, how far is up? We don't know exactly; it may be around 15 billion light-years (at 6 trillion or 6 million million miles per light-year), or it could be larger.

✳

WHY IS THE sky dark at night? This is a question a child could ask, but it was a problem that stumped Kepler and Newton, and baffled Einstein for a while as well. Ever since the emergence of the Copernican model of the solar system, scientists became aware that the solar system was not the universe. Other stars had to be mighty far away, or their motions would have responded to the Earth's motion about the Sun; they were so far away that they had to be other suns, not planets, in order to be seen at all. Kepler and Newton worried about this and they imagined an infinite cosmos. Otherwise gravitational forces acting between stars would pull the stars all together into a big blob. No, the cosmos had to be infinitely big and infinitely old. It had to be static with no cumulative motion, and with stars strewn more or less randomly throughout it. But here lies the rub. If space stretches on forever, every line of sight must eventually intersect a star. Stars would appear to overlap each other with no darkness in between. Every bit of sky in any direction would appear as bright as an equivalent section of the solar surface. Heinrich Olbers, an early-nineteenth-century astronomer, brought this problem to widespread attention, revealing a dilemma in which something was very wrong with this infinite universe, in a concept now known as Olbers's paradox. Olbers's paradox says that in an infinite universe filled with stars, the sky should be a continuous blaze of light, yet the night sky remains dark. Why?

Enter Albert Einstein. His general theory of relativity, published in 1915, relates space and time together as Newton's theories do not. Gravity curves space; the shape of space is affected by the matter in it. No one has a convenient portrait of this universe model, for it cannot be made or imagined. With time as a fourth dimension, this ineffable void may have been described by Casey Stengel about as well as anyone. He maintained, "If you walk backward, you'll find that you can go forward, and people won't know if you're coming or going."

Einstein found to his dismay that relativity predicts a nonstatic universe; it must either expand or contract. He blundered then, by formulating a constant term that forced it to remain constant in size. This, his worst goof, was revealed by Edwin Hubble and others, who found that galaxies are indeed running away from us and each other, and we do live in an expanding universe.

Now Olbers's paradox is explained. Galaxies are distributed sparsely enough that in most directions there are no stars, and the sky is dark. At some distance, galaxies would be moving away at the speed of light and thus could not be seen. Furthermore, the universe is not infinite in time or space. It is about 15 billion years old, and one sees nothing beyond 15 billion light-years.

7

Of Time and the Sky

The sky in the daytime changes very little. It does of course pass from clear to cloudy and back again, and clouds come in seemingly infinite varieties. But the clear sky in the daytime is an empty place, a cerulean dome with one very bright thing up there—the Sun. Once in a while, the Moon also puts in a daytime appearance, and upon rare occasions, the planet Venus barely does so as well. But usually the Sun has it all to himself. Like a colossus he strides across his sky as if it were his own private preserve, as in truth it is. It doesn't take very long even for a child to realize that the presence of the Sun is what makes daylight and differentiates it from night. This did not, however, prevent one military administrative official of an observatory to pose the question: "You say that the Sun is a star, then why can't we see the Sun at night?"

The Sun's march across the sky reflects the two great movements our world undertakes. The rotation about its axis and the revolution about the Sun give us the day and the year, the two fundamental time periods upon which all solar-calendar and time-reckoning schemes are built. A third period, resulting from the Moon orbiting the Earth, is the month, which also plays a part in our marking of time in the Jewish and other lunisolar calendars. All else only amounts to divisions by 12s and 24s and 60s of these periods for hours, minutes, and the like. The week of 7, or in some cases another number of days, is artificial and harks back to astrologically dominant times.

One vital difference between our spin about our axis and our dance about the Sun must always be kept in mind. This is the fact that the two do not move in the same plane. Instead the plane of each is inclined with respect to the other by an angle of 23½°. This angle goes by the names of inclination, obliquity, or tilt, but by whichever name, that angle is the major factor in giving us our seasons.

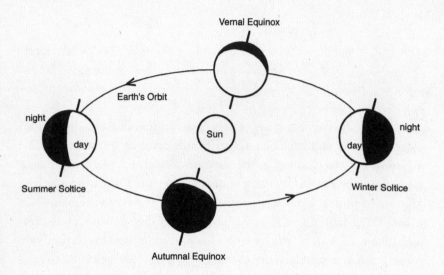

7.1   The tilt of the Earth's axis and the seasons

Strictly speaking, the seasons result from one major and two minor causes. The dominating factor is the one most of us were taught in school. Most globes are built to show a cant or tilt, by not being "upright"; that is, the axis of the globe leans over like the tower at Pisa, only more so. Globe makers do know how to make a model that does not lean over, but they choose to include the lean, thus making science teachers think before they speak.

The reason for the leaning globes is the inclination of the Earth's axis with respect to the direction perpendicular to its orbital plane, which astronomers call the plane of the ecliptic. This angle of 23½° (varying slightly over the millennia) tilts the Northern Hemisphere toward the Sun and the Southern Hemisphere away from it in June. In December, the positions are reversed. Taken together with the direction of the axis with respect to the direction toward the Sun and the latitude of the observer, the tilt determines the maximum angle of incidence of the solar radiation (that is, the altitude of the Sun above the horizon at noon), as well as the duration of daylight at a given latitude. In Australia, or "down under," the seasons are reversed. Christmas can be hot, with sunlight lasting far into the evening, while a displaced American may celebrate Independence Day in the snow.

It makes perfect sense to mark time by the motions of the Sun; after all, the energy source for the Earth and all life upon it is the Sun. Just an average sort of star and gaseous throughout, it is a furnace made of its own fuel. But with 99.8 percent of the mass of the entire solar system, and the next closest star about 268,000 times as far away from us, the Sun is king around here.

Fortunately for us, the radiation emanating from the Sun is very nearly constant and has been so for millions of years. The amount of heat falling on the Earth's surface depends on the angle of incidence of incoming sunlight and on the length of the period of daylight. It reaches a maximum when the Sun is most nearly overhead and becomes insignificant for very large angles near sunrise or sunset. In addition to the diurnal variation of the altitude of the Sun in the sky, a pronounced seasonal variation is present in the incoming solar radiation, at least for high and intermediate geographical latitudes. For examples of this, the maximum length of sunshine at New York at latitude 41° is 3 minutes under 15 hours (on June 22) and in December it is 24 minus this amount, or about 3 minutes over 9 hours. At London, latitude 51½°, the longest and shortest periods of daylight are 16 hours and 36 minutes, and 7 hours and 24 minutes. (But in every case the true day is actually about 4 minutes longer due to atmospheric refraction.)

This is all a natural consequence of the formation of the Earth condensing as it did into a ball that revolves around the Sun in the plane of the ecliptic, but rotates in a plane inclined to it by that 23½° angle.

Stated another way, this plane is inclined by 23½° from the plane of the equator extended onto the celestial sphere, forming the celestial equator.

The ecliptic defining the path of the Sun in the sky crosses the celestial equator at two points. The Sun, then, must cross the celestial equator twice each year. It does so on or very close to March 21 and again on September 23 and at these times the periods of daylight and darkness are each 12 hours long as seen from any point on the Earth. These points and dates are known as the equinoxes, the vernal equinox and the autumnal equinox, respectively. After March 21 the Sun moves northward away from the celestial equator until June 21, when it reaches the most northerly point on the ecliptic, called the summer solstice. The Sun then moves south for 6 months, passing through the autumnal equinox, until it reaches the winter solstice about December 22. The four key dates, the two equinoxes and the two solstices together, are called the colures, and the passages of the Sun define the astronomical seasons.

In designating these points, we are clearly being Northern-Hemisphere chauvinistic; spring, after all, comes in September to lands south of the equator. But most of the land and nearly 90 percent of all people live north of the equator and inhabit the Northern Hemisphere. Hence we will define them as we do above, in the traditional manner.

✳

OTHER ASTRONOMICAL FACTORS have a small but real influence on the seasonal effect. One is related to the distance between the Earth and the Sun. In 1609, Johannes Kepler published his discovery that planetary orbits are elliptical and not circular, with the Sun not at the center but at one of the two foci of the ellipse. Ever since that time the Sun's distance has been known to vary from the average of about 93 million miles or 150 million kilometers. The separation reaches a minimum in early January and a maximum six months later, in early July. The closest and farthest points along the planet's orbit are called perihelion and aphelion, and differ by about 1.5 million miles, 2.5 million kilometers, on either side of the average distance. In these extreme positions the Earth receives about 3 percent more and less solar energy in January and July, respectively, than it does on average. At present the arrival of the Earth at perihelion happens in the wintertime in the Northern Hemisphere and summertime in the Southern Hemisphere, while the farthest point, the aphelion, is reached when the seasons are reversed. This coincidence has the effect of intensifying the seasonal extremes in the Southern Hemisphere while moderating them in the Northern Hemisphere. Since most of the Southern Hemisphere is covered by water (unlike the Northern Hemisphere, dominated by land, as is apparent on any globe), the intensification there is not very pronounced, because the thermal inertia of the oceans tempers the climate; that is, the oceans heat up and cool off more slowly than the land. This feature of maritime climates causes them to be characteristically less extreme in temperature than their continental counterparts. The variation of distance from the Sun, and the sizes and distribution of the continents and the oceans, are the two minor causes of the seasons.

The times of perihelion and aphelion passages have not always taken place in January and July, as they do now. Instead they move ever so slowly, at a snail's pace, around the sky. The reason for it comes from another, much slower motion, but one that is just as important as the day, month, and year. This is the precession of the equinoxes, usually referred to as simply the precession. It is a gyroscopic motion of the Earth's axis bringing about a wobble with a period of 25,800 years. This slow, steady wobble is and defines the precession and was first discovered by the Alexandrian astronomer Hipparchus, in the second century B.C. (a remarkable achievement, given the degree of precision necessary to detect the very slow motions of the stars that it imposes), although it remained for Sir Isaac Newton in the late seventeenth century to explain the reason for it. Precession is the result of the gravitational influence of the Moon and the Sun on the equatorial bulge of rotating oblate Earth. Newton realized it to be identical to the motion of a spinning top or gyroscope as its axis starts to lean away from a vertical orientation. The Earth's gravity wants to pull the top over, but it responds by precessing to the side. For the same reason, the axis also moves sideways. Many aspects of the sky, including the orientations of the constellations, change slowly over the centuries due to the precession.

As a result of the precession, the Earth will be tilted—its axis will be pointing—in the opposite direction 13,000 years from now from its location today near Polaris, the Pole Star, at the present time. This in turn will enhance the extremes of the seasons in the Northern Hemisphere and moderate them south of the equator, just the opposite of the case today. The great land masses of the Northern Hemisphere—Eurasia, North America, and most of Africa—do not share the gentler maritime climates of their smaller southern counterparts (excluding uninhabited Antarctica) and the seasonal extremes in the north are expected to be somewhat greater than at present.

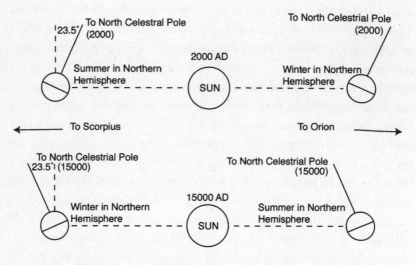

7.2   The inclination of the Earth's axis now and in 13,000 years

As the equinoxes and solstices are carried backward (westward) along the ecliptic by the precessional motion, the passage of the Earth through the closest and farthest points from the Sun will advance in the calendar. In fact, in the middle of the thirteenth century, not long after the Magna Carta was created and Genghis Khan was terrorizing much of Eurasia, these two points coincided with the two solstices. At that time, the closest point and the winter solstice (when the Sun is farthest south) fell together, about December 22, and their opposite numbers came together on June 22 amid the summer season. Now they are separated by about 14 days; each pair of nearby points has separated at the rate of a day every 50 years or so. In roughly 21,000 years, the two solstices will wheel around the whole sky and return to their present positions. Although the true period of the precession is, as we said, near 26,000 years, its effect on climate acts over a rather shorter one closer to 21,000 years. The reason for the difference lies in the superposition of its true period with two longer periods of variations in our orbit.

Other planets have their own seasons. Curiously, all of the planets except Jupiter and Uranus have axial inclinations about the same as ours—about 20° to 30°. Jupiter shows one extreme with a tilt of only 3° while Uranus exemplifies the other at 81°. The seasons on Jupiter due to the axial tilt would be nonexistent, and those of Uranus would be in the extreme. Were it not for the other astronomical influence—the orbital eccentricity—Jupiter would have no seasons at all. Our figure of 1.6 percent for the Earth's orbital eccentricity gives us the third most nearly circular orbit among the planets; only Venus and Neptune have rounder orbits. The eccentricities of the orbits of Jupiter, Saturn, and Uranus are near 5 percent, that of Mars and Mercury, about 9 and 20 percent, respectively, and Pluto leads with 25 percent. So Jupiter has seasons after all. Unlike the Earth, its two hemispheres would be in phase—both experience summer and winter together when it is near its perihelion and aphelion points.

✳

THE NIGHTTIME SKY is a very different affair from the daytime sky, particularly when it is clear. Many feel that there is no grander sight in all of nature than the heavens on a moonless night. Here the stars take over; the Moon can indeed make its mark, as can the five bright naked-eye planets to a lesser degree. With the Sun and Moon, these planets—Mercury, Venus, Mars, Jupiter, and Saturn—constitute the seven visible members of the solar system in the sky—the same five planets that, again with the Sun and Moon, give to the number seven its special significance in the paranormal world, that sphere of astrology, mysticism, Tarot cards, and all the rest. Then Copernicus detailed an alternate system with an eighth body, our own Earth orbiting the Sun between Venus and Mars, and astronomy parted company with astrology soon afterward.

Prior to Copernicus, since the times of Plato and Aristotle, the universe had been thought of as consisting of a model of two spheres. One is our world and it is very small compared to the other, the sky upon which all of the stars are fixed, at the same distance out beyond Saturn, then the farthest known planet. The Sun and planets were in between the two. It is easy to appreciate this two-sphere model as it was known, for no one can see any depth in the sky; everything appears as if at the same distance. And, in fact, we have not thrown over the two-sphere cosmos in every aspect; it is used as the basis for surveying and for celestial navigation. These are activities in which only the direction and its change with time called motion are applied. The radial distance to any object, whether the Moon or the farthest star, is not a factor in these endeavors. To this extent, perception, placing all at one immense distance, simulates reality sufficiently well.

Except for the seven objects that are seen to move across the sky, all the rest of the vault of heaven is taken over by the stars. With the exception of a few very faint nebulae and our solar family, everything visible to the eye is stars; even the Milky Way is nothing but a mass of faint, unresolved stars. It is easy to fathom from a glance at the sky that every civilization and society has divided up the stars into recognizable patterns called constellations, or asterisms if smaller. The great societies of the East favored several hundred small constellations, whereas the Western world tended toward fewer larger ones. The standard number of constellations of the classical world stood at forty-eight; these are still the ones we use. Orion, the two bears, Cassiopeia and her husband, Cepheus, and all twelve signs of the zodiac are among that group of forty-eight. So matters stood until the invention of the telescope and the nearly contemporary travel of Europeans to African regions south of the equator, where the deep southern sky can be seen. Forty others gradually came about to fill in the faint interstices between the more visible forty-eight, and the deep southern sky, eternally invisible to northerners, was in its turn divided into constellations. The major ones have formed a dominant role in our cultural mythos, from the *Iliad* and the *Odyssey,* through the

Divine Comedy and the works of Shakespeare, and directly into modern culture. The dramaturgy of the night sky is with us still.

How long will our familiar constellations last? How many thousands of years must pass before all of our star groups are so distorted by the individual proper motions of their member stars that no one living today could recognize any of them? Most textbooks intended for courses in descriptive astronomy limit themselves to a brief mention of proper motion and may include an illustration of Ursa Major, the great bear, whose brightest part is known as the Big Dipper or Plough, as it appears today and also 50,000 or 100,000 years in the past. It seems that the recognition of the dipper has endured from before the end of the last ice age and well before recorded history, and is likely to last into the next ice age at least.

Edmond Halley was the first to prove that stars were not fixed for all time but moved across the sky, each with its own individual motion, called proper motion. He discovered this in about 1700, as he was comparing his star positions with those of Hipparchus, the great Alexandrian astronomer of the second century B.C. Hipparchus had discovered the very slow celestial motion that takes longer than a human lifetime to detect by eye. This is the precession, the gyroscopic spin of the Earth's axis that shifts the location of the celestial poles, the equinoxes, and the solstices about the sky, but does not affect the positions of stars relative to each other. Halley had detected that the bright nearby stars Sirius, Procyon, and Arcturus had shifted about 1° with respect to their stellar neighbors in the intervening two millennia.

In 50,000 years, the Big Dipper or Plough will be very distorted and barely recognizable, even though five of its stars share a common proper motion, and even in the 25,800 years or one precessional period that will bring Polaris back around to the Pole, the shape will be noticeably altered. Sirius and our other nearby neighbors will be well away from their present locations by that time, and some of them may appear as nondescript, fainter stars. Others now inconspicuously approaching us may shine more brightly in their places, but a few asterisms may remain recognizable for much longer.

By 100,000 years hence, we can say (from the astrometry of position and proper motion) with some certainty that most of the familiar star-figures we know will be unrecognizable. This is not to say that all of our bright neighbors will be faint and far away. Rather their motions will have scrambled the forms we know. This can be seen in the illustrations following (Figures 7.3–7.5), which show the familiar forms of Ursa Major, Cassiopeia, and Leo today and as they will appear in 100,000 years. Their stars remain bright but their forms are then lost to us.

However, two of our best-known star figures will be with us far longer. Consider the year A.D. 802,701, the year to which H. G. Wells's time traveler, in his noted novel *The Time Machine,* ventured to journey to examine our species as it will have evolved at that time. The narrator of the tale notes that all of the old constellations had gone from the sky, and the stars had long since rearranged themselves in unfamiliar groupings. This is almost true, but two major exceptions will remain. Some simulacrum of today's star patterns will still exist in that distant epoch. Above all will Orion, the great hunter, still bear a likeness to its present self for all but one of its bright stars move nearly together in a group. To be sure, his shoulders will be grotesquely broader, with Betelgeuse, the odd man out, having shifted about 6° to the northeast, but the other bright stars will appear about as they do now. The solar system will have traveled 50 light-years in that time, not ½₀ of their distance.

One other constellation has the staying power of Orion, and that is Scorpius. With all the noble beasts that our forebears could have placed in the sky, it may seem odd that this magnificent group should honor one of the least appealing. But Scorpius is among the few constellations named for a creature it resembles. It, too, may retain a tattered but still perceptible vestige of its former self to the Eloi and Morlocks, or whoever is alive in that remote time, 8,000 centuries hence. Like Orion, it lies near the Milky Way, and most of its bright stars also belong to a group of very young, very luminous stars. Indeed extending across much of the deep southern heavens is the star-rich Scorpius-Centaurus Association, sprawling across those two bright constellations and a few smaller ones. Also like Orion, this is a group of distant stars of common origin moving together and showing little change across the eons.

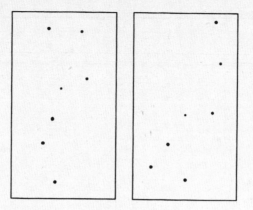

7.3   The Big Dipper as it appears today (*left*) and in 100,000 years (*right*)

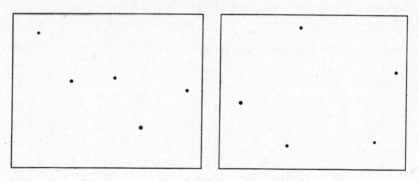

7.4   Cassiopeia as seen today (*left*) and in 100,000 years (*right*)

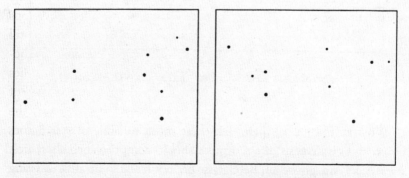

7.5   Leo as seen today (*left*) and in 100,000 years (*right*)

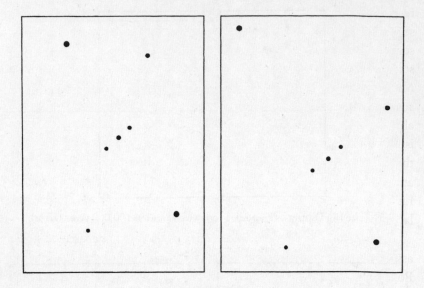

7.6   Orion as seen today (*left*) and in A.D. 802,701 (*right*)

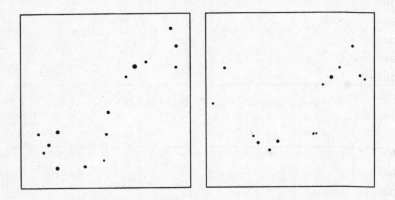

7.7   Scorpius as seen today (*left*) and in A.D. 802,701 (*right*)

What an irony it is, then, that these most indelible of star figures, Orion and his nemesis, the scorpion, should comprise such short-lived stars. Rigel and, above all, Betelgeuse are not likely to be alive as visible stars at that distant epoch. Betelgeuse in its supergiant phase is already terminal, with only hundreds or perhaps thousands of years left to it. The second-magnitude stars of the belt will last longer, but not by much. The scorpion shares this demise as well. At its heart lies its lucida or brightest star, the red Antares, so named because its color makes of it a rival to Mars whenever that red planet is nearby. Antares, like Betelgeuse, is a supergiant with little longer to live. Orion and Scorpius may recognizably mark our skies for a longer time than any of the other familiar asterisms, but with the one bereft of a shoulder and maybe a knee as well, and the other losing its heart. The recognition of their shapes, almost alone among constellations, could end in a unique fashion, astrophysically rather than astrometrically, with some of their brightest stars in the stellar graveyard before they lose their memorable forms.

These two great physically connected groups of stars are called stellar associations and appear as if balanced across a giant fulcrum at the South Pole of the ecliptic, on equal and opposite arms about 70° in length. The arms extend due north in opposite directions. As one rises the other has just set, as we see them from the Northern Hemisphere. Over the centuries, precession will swing them alternately up and down, north and south in a rhythmic fluctuation, like a 13,000-year pendulum, in turn raising one to the celestial equator and burying the other in the deep southern sky. The higher one will be seen among our winter stars and the other will be all but lost in the deep southern haze of summer. They will retain their apparent animosity and distaste for each other with each rising just after the other sets.

We are reminded of their mythology of classical times in which the scorpion was said to slay Orion with his sting, and now faces extinction himself from Sagittarius, the archer, nearby. As long as our axis remains in place or nearly so, these two antagonists, never seen together from our purlieu, will mark the last identifiable celestial hallmark of our era.

The night sky in 8,000 centuries will not be our night sky. It would retain the broad features of today's nocturnal scene, but almost all of the familiar stars and constellations would be gone, or in the past would not yet have come onto the scene. The nearest objects, the members of our solar system, would appear the same, and the farthest, the great Milky Way, faintly arching across the heavens, would also resemble its present self. But hardly one of the individual stars familiar to us could be seen. Most stars are bright simply because they are relatively close to us and they would be lost in the background; the truly intrinsically bright stars, the beacons that shine throughout much of the Galaxy, have short lifetimes. From birth to death, these supergiants are exhausted in far fewer than 65 million years. We know from galactic dynamics that in place of our familiar stars, another set would appear that would possess about the same general properties.

✳

SIXTY-FIVE MILLION years ago, planet Earth took a hit and its lush Mesozoic ecosystem was mortally wounded. The damage was done by an asteroid or possibly a comet perhaps 10 miles in diameter, racing toward the Earth at several tens of thousands of miles per hour. The most widely spread and richest fount of life in history died, along with some three fourths of its species, including almost all the large ones.

What did the sky look like in the days just before that holocaust? We know that the daytime sky looked then just as it does today. The Sun appeared the same size and moved at the same speed across the sky as now, due to our orbiting about it, but the day, the rotation period of our Earth, would have been about half an hour shorter than the 24-hour day we know. Several factors slow down the rotation very slightly. Primarily, the retard is brought on by the Moon and, to a lesser extent, the Sun. They cause the tides, which slosh up onto continental shelves from the deep and act as a brake on rotation, slowing it down. Over the eons the tidal friction will slow our day to a duration of a month or more, but not until billions of years in the future.

PART II

The
Larger
Sky

8

*

Full Moon and Mumpsimus

Mumpsimus.
1. One who obstinately adheres to old ways, in spite of the clearest evidence that they are wrong; an ignorant and bigoted opponent of reform.
2. A traditional custom or notion obstinately adhered to however unreasonable it is shown to be.
3. Stupidly conservative.

Oxford English Dictionary, 2nd edition

Nicholas Sanduleak was an astronomer, a shy and retiring man with a legendary sense of humor. Among his many other research activities, he devoted time to the study of the authenticity of astrological predictions and other claims of a dubious nature. One of the claims he investigated not long before he died, one of the most widely known and believed among all things associated with the sky, is the popular allegation that crime, insanity, and other aberrant human behavior increase around the time of Full Moon.

Nick and I were graduate students in astronomy together at Case Western Reserve University in Cleveland, and he remained a close life-long friend. His study of the frequency of crimes (mostly of passion) in Cleveland was published in the *Skeptical Inquirer* in 1985, under the title, "The Moon Is Acquitted of Murder in Cleveland." He refuted earlier conclusions bolstering common suppositions that these crimes increased whenever the Moon presents all of its sunlit hemisphere to the Earth.

Throughout the whole of recorded history, the Full Moon has been believed to have a pernicious and unwholesome influence on humans, animals, and those who might fall in between. Despite the beneficial influence on poets and songwriters who pursue their muses at this phase, the conviction that the time of the monthly cycle when the Moon is full increases nefarious and erratic behavior is more firmly believed than the facts that most weather forecasts are correct and most Americans are not obese. This may well be the most widely held of all erroneous notions that involve the night sky. The word *mumpsimus,* defined above, applies here if anywhere.

Nick felt the need to discredit indefensible assumptions about the phases of the Moon. One such conclusion was published by Arnold Lieber in his book *The Lunar Effect.* Nick quotes Lieber as maintaining that "the Moon is able to adversely affect the mental and emotional stability of humans by means of raising physiologically disruptive 'biological tides' in our bodies akin to the tides it raises in the Earth's oceans." Since the human body is largely composed of water, the analogy seemed plausible to many.

Lieber quotes a prior investigation he made into the incidence of homicidal assaults in Miami and Cleveland, in which he reported them to rise with the Moon at both new and full phases over other times; at either phase, the amplitude between high and low ocean tides rises to a maximum due to the combined lunar and solar action. High tides are at their highest and low tides are at their lowest whenever the Moon and the Sun are at syzygy; that is, whenever they lie along or nearly along a straight line with the Earth. Some other investigations have corroborated, and others have refuted, an increase of criminal events at or near the Full Moon.

A number of investigators have found a problem of perception here. "Every policeman, bartender, and emergency room attendant," Nick explains, "*knows* that people tend to act crazy and become more violent when the Moon is full." No comparable extremes in behavior get reported at the New Moon, a fact that discredits the tidal theory reported by Lieber and others.

No, it is at the Full Moon when bizarre and deviant actions are perceived to reach a crescendo. This belief goes back for centuries in many cultures. Notable are the legends of werewolves and vampires and other things that go bump in the night. Hollywood has planted and reinforced this elusive reality into all of us, through the many films starring Doctor Frankenstein and his redoubtable monster; Count Dracula as a latterday incarnation of the notorious Vlad, the Impaler; and the Wolf Man, a normal person until the Full Moon rises, at which time he is transmogrified into a hirsute werewolf and all-around fiend. Boris Karloff, Bela Lugosi, and Lon Chaney Jr., respectively, made their fortunes playing these ghouls.

Movies of this genre required and took poetic and artistic license with the Moon, by making it full for days or weeks on end. These actors' names on the marquee guaranteed a full Moon in every nocturnal scene, always through a predictable veil of haze or mist or broken clouds, usually above a swampy miasma. No night for flights to the Moon, full or otherwise, on gossamer wings, as Cole Porter writes; for that, we require a clear night devoid of excessive water vapor in all forms.

But how long does a full Moon really last? For how many consecutive nights can we say the Moon is full? In astronomical reality, the time of Full Moon is but an instant, the exact moment when the Moon is opposite the Sun in the sky in measure in the east-west direction, called right ascension, and analogous to longitude on the Earth. Most passages across this line in the sky find the Moon orbiting either to the north or to the south of the Earth's shadow, which is always centered at the point directly opposite the Sun in the sky. If it passes very close, close enough to pass into the shadow, we observe a lunar eclipse, because the Earth cuts off the sunlight and the Moon is nearly dark. (A little light is refracted through our atmosphere into the dark shadow; hence, the Moon remains illuminated by a faint reddish glow.)

But when we hear Debussy's *Clair de Lune* or Glenn Miller's *Moonlight Serenade,* it is always a round soft gentle full Moon that comes to mind. In truth most of us perceive the Moon as round for periods before and after the exact moment of full. I have asked students in my courses to call the lunar phase as they see it each night in order to obtain a consensus on just this point. I have found that the Moon appears full for about 2 days on either side of the moment itself. Further than that, the Moon appears gibbous, the phase occurring both before and after the full phase, when it appears as a football, something less than round. If this is representative of the entire population, we can say that for 4 days every month the Moon appears full.

The interval between one full Moon and the next is about 29.5 days. Thus we can hypothesize that 4 divided by 29.5, or just under 14 percent of the time, the Moon is seen or perceived as full. Is the average number of incidents of violence or craziness during that 14 percent of the month equal to or greater (or less) than the rest of the time? That is the number that Nick sought from police records in Cleveland.

Nick's analysis is much more extensive than I have reported here, but his conclusions are not. He notes that "this study found no evidence that the frequency of homicidal attacks in Cuyahoga County (metropolitan Cleveland), Ohio, during 1971–1981, the years covered in this study, was related in any way to the phases of the Moon or the action of lunisolar tidal forces." He cites a number of other surveys made elsewhere that draw the same conclusions.

"How then does one account for the anecdotal testimony so readily provided by police, bartenders and maternity-ward nurses?" Nick proposes, as have others before him, that hearsay plays a significant role. On a busy or stressful day, anyone responding to these troubles may state that with all of this increased mayhem, "There must be a full Moon." No one troubles to check on this, indeed very few would know how to do so, or to recall the lunar phase at the time. But the recollection is reinforced subsequently, and too often erroneously, that the Moon was full.

This is a reinforcement of a belief perpetuated in the media and among people in general. It is a clear case of mumpsimus, a "persistent belief in the face of conclusive contradictory evidence." For some reason, many want to continue to believe in this persistent and frequently negative influence. It provides comfort perhaps in the same manner that astrology provides comfort, reassurance that the heavenly bodies do constitute visible deities or deifications moving in an orderly universe.

Perhaps the human mind needs relief from a steady diet of rationality. No less a scientist than Sir Isaac Newton may have felt this seminal urge when he took up alchemy for a time. Richard S. Westfall, in *Never at Rest,* his scholarly biography of the great scientist, raises just this point, stating:

> The concept of a secret knowledge for a select few aside, all of the above characteristics (to glorify God, to teach a man how to live well, and to be charitably disposed toward neighbors, etc.) applied as well to the mechanical philosophy, which Newton had recently embraced. In the nature of the truth they offered, however, the two philosophies differed profoundly. In the mechanical philosophy, Newton had found an approach to nature which radically separated body and spirit, eliminated spirit from the operations of nature, and explained those operations solely by the mechanical necessity of particles of matter in motion. Alchemy, in contrast, offered the quintessential embodiment of all that the mechanical philosophy rejected. It looked upon nature as life instead of machine, explaining phenomena by the activating agency of spirit.

Escape from the rational explicable world is indulged by almost everyone in one form or another. Setting aside the ineluctable laws of physics, to embrace an avowed miracle or any teliological suspension of the ethical takes many forms but seems to satisfy a very basic and human need. We see it repeatedly in the color cartoons that used to accompany full-length movies, back when most movies were shorter in time than they are today. Whenever Tom the cat chased Jerry the mouse, or Sylvester chased Tweetie Bird, off a cliff, they paused to look down before falling, one of many cases of the plausible impossible that sustains this genre.

Just what are the influences of the Moon upon the Earth and all life on it? The obvious effect is due to its gravitational pull, and this is most evident in the tides. Although tides occur in our atmosphere, here at the bottom of the ocean of air they are not very noticeable. The solid portion of the Earth, being rigid—unlike air and water—is also pulled on by the

Moon, but the effect on the solid Earth is also all but negligible. It is in the watery part of our sphere that the tides are apparent. The most significant point to keep in mind about the tides is that there are not one but two high tides per day, actually per 24 hours and 50 extra minutes, with two low tides in between. Thus the average interval between one high tide and the next, or one low tide and the next, is about 12 hours and 25 minutes.

Sir Isaac Newton's universal law of gravitation shows the force of gravity to depend upon the inverse square of the distance over which it acts. Now the rim of the world closest to the Moon, the point that would see the Moon at the zenith, lies about 4,000 miles closer to the Moon than does the center of the Earth. But the rim turned away from the Moon is also 4,000 miles from the center but farther away on the opposite side. The Moon pulls most on the closest part of the Earth, next at the center, and least on the far side. Thus there are always two tidal bulges in the oceans, one toward and the other away from the direction of the Moon.

The Sun also raises considerable tides here, but the Moon's effect is the stronger, being about 2.2 times the gravitational tug of the Sun on average, with minor variations according to the varying distances each undergoes in the course of motion in the elliptical orbits of both the Earth and the Moon. When the Moon is at the new phase or at the full phase, it is in line with the Sun and the two act in concert, with the resulting tidal amplitude between high and low tide being greater than the mean amplitude. We call these maxima, these highest and lowest tides, spring tides. But in between, when the Moon is at the first or last quarter phase and appears half illuminated to us, the Sun counteracts the Moon and the tides are weaker, being known as neap tides. At those times the high tides are not as high, and the low tides not as low. Both extremes are illustrated in Figure 8.1 for the two cases of spring and neap tides. The lunar influence is thus gravitational and acts always along the direction toward the Moon itself.

The planets and every other object in the universe cause tides, but they lack the proximity of the Moon and the great mass of the Sun and are thus of little consequence. Among them Venus and Jupiter raise the largest tides because the one is close to us and the other is big and massive, but at best, when Venus is closest, its pull is but $1/8{,}000$ that of the

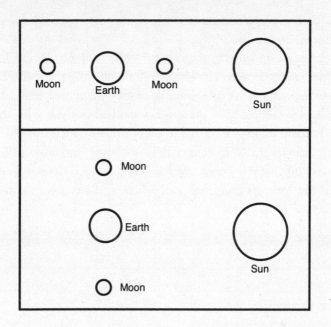

8.1 The positions of the Earth, the Moon, and the Sun at spring and neap tidal conditions. The Moon can be at either new or full to produce the extreme spring tides and at either quarter to produce the less extreme neap tides.

Sun, which has itself only 45 percent that of the Moon. Jupiter's best is only ¹⁄₆₅,₀₀₀ as much as the Sun, and, in descending order, Mars follows at ¹⁄₁₅₀,₀₀₀, and then Mercury and Saturn, both much weaker still. In no way could their tidal influence be detectable on our seas.

What about the variable light from the Moon; this does reach a sharp maximum at the full phase. But moonlight is merely reflected sunlight. If some common folk become werewolves or other kinds of fiends in the amplified moonlight, they would really go ape in the daytime when the light received from the Sun is around half a million times as intense, and this they do not do in the myth of any culture. Furthermore, in our present society almost any nearby streetlight or porch light outshines the Full Moon, often by many times its light intensity. Thus neither the observational data nor any plausible theory can account for this very common belief. No known force from the Moon can drive people crazy at one time more than another.

Nick has most assuredly acquitted the Moon of murder in Cleveland, and others have done so elsewhere in a convincing manner. His study and conclusions are but one example among many that have subjected the Full Moon effect to a thorough statistical analysis with the same result. Is harm done in a refusal to accept these conclusions? I suspect that it is whenever that ignorance may influence legislation or discriminate for or against one person over another. At such times, scientists are responsible for exposing unsound beliefs and practices as my friend Nick Sanduleak has done.

9

✳

The Amazing Analemma

Those who will not reason perish in the act:
Those who will not act perish for that reason.
W. H. Auden

Have you ever glanced at an old globe, maybe twenty or more years old? If so, you may have noticed a large lazy figure eight sprawling somewhere in the eastern Pacific Ocean away from any land. Globes are fairly easy to date, at least to a decade or two; one might look first at Poland, since the post-1945 Poland is differently shaped and displaced westward from that of the earlier model. Before 1919, there was no Poland at all. East Pakistan became Bangladesh in 1971, and Yugoslavia fissioned into five countries in 1991, about the time Slobodan Milosevic lost the first and maybe the second of the four wars he has started. In 1990, East Germany disappeared from globes as a separate entity, and sometime before then, with a few exceptions, this lazy figure eight did the same.

The name of this lazy figure eight is the analemma and it is shown in Figure 9.1.

The common reason given for its disappearance from most recent globes is a sad one. Globe makers claim that the latest round of science teachers in secondary schools cannot explain it to their students, who just don't understand it, and the dumbing of students is nowhere more pronounced than in the sciences. To paraphrase the dumbing drift at its worst, as many parents and teachers of my acquaintance do, why should

9.1   The analemma as
it appears on a globe

students in America learn about the analemma, as long as they keep up
their soccer, jogging, and rock climbing?

The analemma marks not only the seasons but also the two principal
reasons for them. As reported earlier, the first and more significant is the
inclination or tilt of the axis with the axis of its orbit around the Sun, an
amount of 23½°. The lesser cause is the fact that the orbits of planets
about the Sun are ellipses, not circles, with the Sun at a focus, not at the
center. Our orbit is not far out of round; at its closest and farthest, the
Sun can be about 91½ million or 94½ million miles away, respectively;
about 1.6 percent nearer and farther than the well-known average of 93
million miles. The points of closest approach and farthest distance are
known as the perihelion and aphelion points. These two factors, the tilt
of the axis of rotation and the eccentricity of the Earth's elliptical orbit,
not only cause the seasons, they also cause the Sun to appear to move
along the ecliptic around the sky at an uneven rate.

At the present time, the Sun appears farthest north at the summer solstice, about June 21, as we know, and its most southerly point, the winter solstice, is reached about December 22. The Gregorian calendar established by Pope Gregory XIII in 1582 is arranged to keep these dates fixed, along with the vernal and autumnal equinoxes that occur on or about March 21 and September 23, respectively. By definition, the equinoxes and the solstices lie on the ecliptic, the apparent path of the Sun around the sky, and the equinoxes also lie on the celestial equator and mark the two points where it and the ecliptic intersect.

The date of the Earth's arrival at the point of perihelion falls within a day of January 5, and aphelion near July 5, as mentioned earlier. But over the centuries these dates would be expected to move forward in our calendar, by about 1 day every 72 years, or once around the sky each 25,800 years. This happens because of the precession, described earlier. The true values are closer to 58 years and 21,000 years, due to the combination of the precession with motions of an even longer period.

At present, the dates of the apsides (a collective name for the perihelion and aphelion points considered together) line up closely with the solstices. Thus our closest approach to the Sun falls only about 2 weeks after it reaches the most southerly point, the winter solstice. Similarly, we are farthest just after we pass the summer solstice. We attain distances from the Sun of about 1½ million miles off the average distance. This makes for slightly cooler summers and warmer winters in the Northern Hemisphere than would hold for a circular orbit, with more extreme seasons in the Southern Hemisphere.

But, as discussed earlier, the slow wobble we call precession moves the apsides forward. They will shift the equinoxes and the solstices once around in less than 25,800 years, the period of the precession, as they have a much slower motion of their own. So in about 13,000 years or less, it will be the Northern Hemisphere with the temperature extremes and the Southern Hemisphere with the more moderate climate.

The inclination of the axis and the eccentricity of our elliptical orbit not only cause the seasons, but these same two features cause the Sun to be a poor timekeeper. Both force the Sun to run a little ahead or behind a clock keeping exact time, but averaging out over the course of a full year. The sundial, which records the time by a shadow in the sunlight, will be seen to accumulate errors up to as much as about 15 minutes ahead of the clock in early November, and that much behind in early February.

This difference is known as the equation of time, and it is defined as the apparent solar time (as shown by a sundial) minus the mean solar time (as kept by a properly running clock). Thus

$$e.t. = a.s.t. - m.s.t.$$

but both times apply only to the site of the sundial. The analemma tells us at a glance how to convert time by a sundial into the correct local time. To convert into legal time, such as Eastern Standard Time, one must know the longitude of the sundial.

The analemma is another way of graphing this combined relation. In it the vertical axis (the long dimension of the figure eight) gives the position of the Sun in the sky along the north-south direction, called declination. The declination of the Sun reaches from $23\frac{1}{2}°$ north of the celestial equator (the extension of the equatorial plane into the sky) to $23\frac{1}{2}°$ south of it. These two extremes are the solstices, and the equinoxes are located on the celestial equator at $0°$ declination.

In analemma form, we can see the equation of time due to the tilt, the eccentricity, and the combination of the two effects. This last duplicates the earlier representation as well as the figure that appeared on globes, centered on the equator, usually in the eastern Pacific Ocean.

The shape of the analemma is not set for all time. Three long-term variations in the Earth's orientation and motion vary the shape. These are the Milankovic cycles, named for the Serbian astronomer who first recognized that they and the harmonics between them vary the Earth's climate and are mostly responsible for the ice ages. One is the familiar precession, moving the colures westward around the sky every 25,800 years. This shifts their positions with respect to the apsides, themselves rotating more slowly eastward around the sky about once in 100,000 years. The combined period of the two is shorter than either, being about

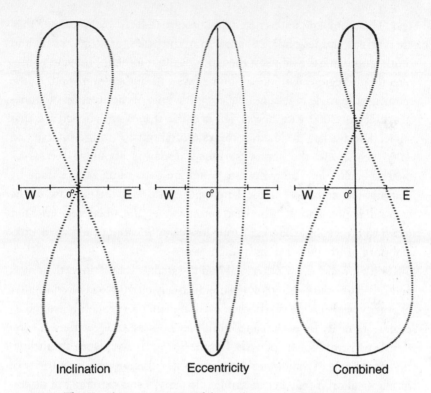

Inclination   Eccentricity   Combined

9.2   The separate components of the analemma and their combination

21,000 years. Thus the apsides move 1° with respect to the colures once in 58 years and were lined up with the solstices about A.D. 1246. At that time we were closest to the Sun on December 22, the date of the winter solstice itself, and aphelion fell on the day of the summer solstice.

The third motion produces a periodic variation in the inclination of the axis, now amounting to 23°26′. The period is about 41,000 years and varies the tilt between about 22° and 24.2°, although the extremes can vary. The last maximum inclination amounted to 24.2° and occurred about 9,500 years ago, and a minimum of 22.6° will take place in another 10,200 years. By definition, the tropics of Cancer and Capricorn lie at this angle north and south of the equator, and represent the limits to the torrid zone, from which the Sun can be seen at the zenith sometime each year. The arctic and antarctic circles also are defined by this angle; they are now at latitudes of 23°26' away from the poles, or 66°34' north and south, and mark the limits of the lands of the midnight Sun, and consequently the regions with one or more days without any sun at all. At the moment, the angle is getting smaller; at the time of the erection of Stonehenge about 2000 B.C., it was 24° and the stones are aligned for that angle. The tilt has declined at the rate of about ½° in the intervening 4,000 years, and the annual shrinkage today is about 48 feet (14.7 meters) each year. The temperate zones are expanding at each edge by that amount annually, almost 1 mile per century at the expense of the tropical region and also by that amount over the polar regions—not much to notice, but the torrid zone loses an area about the size of Luxembourg or Rhode Island every eight years.

The Sun and every other planet and satellite rotate upon their axes and revolve about the Sun or a planet (actually all orbit a common center of gravity between the two bodies; even the Sun moves slightly around its center of gravity for each planet). But they possess a wide variety of rotational and revolutionary periods. For the Sun and each planet, the inclination of the axis of rotation to the plane of its orbit and the eccentricity of its elliptical orbit together determine the length and extremity of its seasons. The speed of its rotation and its density and internal makeup also fix its degree of oblateness. Jupiter and Saturn, with their days of about 10 hours, are the most oblate. Venus has the longest day known; it is about 243 days, about 8 months. As a result the oblateness of Venus is about nil and the planet is spherical. All others lie between these extremes.

Eccentricity of orbit varies, too, among the planets. Recall that Venus and Neptune have the roundest orbits with eccentricities not much in excess of 0.5 percent, followed by Earth at 1.6 percent. All of the others are larger, with Jupiter, Saturn, and Uranus near 5 percent, Mars at 9 percent, and Mercury at 20 percent. Among the planets, Pluto runs to 25 percent, but any comet has this beat; Halley's Comet is about 95 percent with a perihelion distance near Mercury and an aphelion beyond that of Neptune.

Most of the planets have axis tilts not far from ours. But two stand out; Jupiter's is only about 3° and Uranus's is as high as 81°. On Jupiter, the seasons, such as they are on that cloud-enshrouded giant, are almost solely caused by its orbital eccentricity, and the two hemispheres would be in phase, unlike on the Earth.

Uranus has the strangest seasons of all. Imagine the Sun in our sky, moving north until it is close to the Pole Star, casting almost the entire Northern Hemisphere into sunlight all day and all night. Then swiftly the Sun would shoot southward, leaving us in total darkness for months. Intense heat followed by an endless permeating cold would be the lot of the whole planet year after year.

The illustration below depicts a very familiar diagram, showing, as it does, the track made by the north celestial pole every 25,800 years due to precession. But in this diagram, the approximate present positions of the

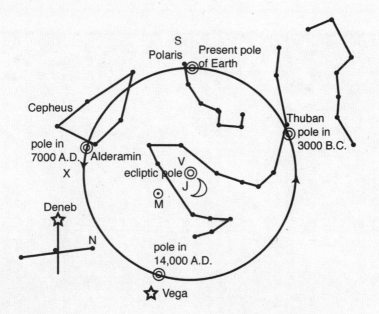

9.3   North poles of the Sun and Moon and most of the planets and our precession. The symbols are defined as follows: ⊙ = Sun, ☽ = Moon, M, V, X, J, S, and N represent the north poles of Mercury, Venus, Mars, Jupiter, Saturn, and Neptune. Uranus and Pluto are not represented because their poles lie well outside this region.

north poles of the Sun and most of the planets are also shown. Jupiter's, as expected, is close to the north ecliptic pole, the point perpendicular to the plane of the ecliptic, analogous to the north celestial pole's relation to the equator. Mercury, Venus, and the Sun line up like Jupiter in that their tilt is small, only a few degrees. That of the Moon is also, but wanders about by a few degrees. The poles of Mars and Neptune line up near the star Deneb as indicated, and Saturn's pole is not far from Polaris. With their huge tilts, Uranus and Pluto have north poles near our celestial equator, near the constellations of Libra and Aquarius, respectively.

The analemma has been photographed. Astronomer Dennis di Cicco produced the famous picture reproduced here and first published it in *Sky & Telescope* magazine. Di Cicco fixed a camera in place and took a

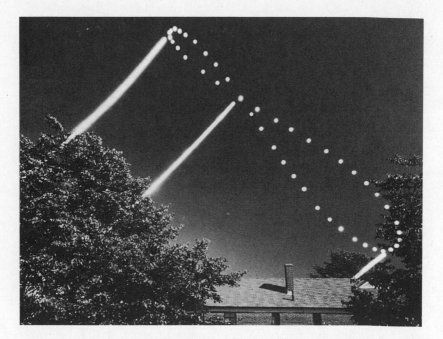

9.4 The analemma as photographed by astronomer Dennis di Cicco. The dots represent forty-five exposures of the solar disk, taken about 8 days apart, each at 8:30 A.M. EST. Together they trace the analemma in the sky. Photo by Dennis di Cicco, Sky Publishing Group

photograph of the Sun every 10 days or so throughout the year, at pre-cisely the same time of day. Since the mean solar time was the same for all exposures, the apparent solar time was not. Sometimes the Sun was ahead of the clock and sometimes it was behind. This, and the normal seasonal motion of the Sun north and south in the sky, produced a lazy figure eight of precisely the shape of the analemma appearing on globes. He then exposed the shots for a longer time to get the foreground in the picture. The photograph took a year to complete, and is now justifiably famous.

Almost-Round Earth

From the Observatory, I got a glorious view of the surrounding countryside. To the north lay the Tirolean Alps, snow-capped and almost hidden in clouds, and joined in the north-west by the hills of Vicenza. In the west and nearer, I could make out distinctly the folded shape of the hills of Este. To the south-east an unbroken plain stretched away like a green sea, tree after tree, bush after bush, plantation after plantation, and, peeping out of this green, innumerable white houses, villas and churches. The Campanile of San Marco and other lesser towers of Venice were clearly visible on the horizon.

Johann Wolfgang von Goethe, *Italian Journey*

When I read this excerpt from Goethe's journal of an extended tour of Italy in 1786–1788, I was surprised at the statement that he could see from Padua the campanile of San Marco at the center of Venice and other structures there. Goethe mentions telescopes elsewhere in the book and he may have used one here.

Views of distant cities and their buildings are not rare in literature; in Thomas Hardy's novel *Jude the Obscure,* young Jude glimpses from a hill "the whole northern semicircle between east and west, a distance of forty or fifty miles." After some time, Jude saw shining spots of reflected sunlight on the structures of Oxford, called Christminster in the book. Poor Jude did not have a telescope and likely could not resolve these reflections further.

Are these two accounts true or realistic? How far can one see on a clear day? Padua and Venice lie 22 miles apart, and as for Christminster (Oxford), it lay "near a score of miles from here," from the spot that Jude searched for it. Since the Earth is round, the limiting distance under very clear conditions is fixed by its curvature.

We can easily determine the distance to the horizon, given only our height above the general surface, or sea level if appropriate. Hardy sets Jude on a hill, and Goethe speaks of an observatory, also likely to be on a hill. Since both see a number of buildings, not just the tips of the tallest, they viewed a horizon well below the top of the skyline. We can set up a diagram such as Figure 10.1, in which a triangle appears whose apices are at the location of the observer, the objects seen afar, and the center of the Earth.

This is a right triangle with the right angle at Venice or Oxford (in these cases). Pythagoras in the sixth century B.C. established his well-known rule that the square of the hypotenuse of such a triangle equals the sum of the squares of its two sides. In this case, the latter are d, the distance to the object or horizon, and R, the radius of the Earth. The hypotenuse is R + h, where h is the height of the viewer. Thus we have

$$(R + h)^2 = R^2 + d^2$$

If either d or h is known, the other can be found.

10.1 Variation of maximum distance with height. The triangle shows the relation between the radius of the Earth, R, the height of a viewpoint above the surface, h, and the distance to the horizon, d.

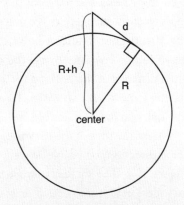

Our planet's radius, R, is about 3,957 miles in length. In order to see for a distance of 20 miles, a quick calculation on a pocket calculator shows that R + h is 3,957.0505 miles from the center, making h to be 267 feet, not a great height for this purpose. But Jude could see 40 to 50 miles in all directions. This requires the unlikely heights of 1,067 and 1,668 feet, respectively, and well beyond the height of any tower of the time or hill at that location, since the heights must increase very rapidly and disproportionately with a modest increase in the distance seen. The top of the Empire State Building is 1,250 feet above the streets of New York and 200 feet above the tip of the spire on the Chrysler Building nearby. No office building is taller than the 1,509-foot World Financial Center, now under construction in Shanghai. To see 40 or 50 miles was not possible in Hardy's day unless one was looking at mountains or standing on one at the time.

An airplane on a commercial flight flies normally at about 33,000 feet (10 kilometers). Here, for all its height, we can see only about 222 miles out to the horizon. The dip in that horizon is substantial, being near 3.2° below the horizontal plane, but the horizon, seen from more than twenty-five times the height of the Empire State Building, is only about five times as far off. The height must increase ever more rapidly for the distance and dip of the horizon to proceed at a constant rate.

One very clear smogless day in New York, I went to the top of the Empire State Building with my binoculars to see as far as possible along the Connecticut shoreline, with which I am very familiar. A stiff breeze blew any smoke and haze out to sea. I was just able to spot a smokestack in downtown Bridgeport, 47 miles away, from the higher observatory on the 102nd floor. Nothing beyond the smokestack was visible. The distance to it is consistent with the case illustrated above, since the calculated distance of the horizon for 1,250 feet is 43 miles, but the building's base is some 50 feet above sea level and the smokestack is more than 300 feet tall, making it accessible at that distance.

Might the tops of the towers of New Haven, 16 miles farther on in the same direction, be visible as well? For this we use the relation shown in Figure 10.2. The smaller triangle to the right shows the situation for d = 16 miles; here h must be at least 171 feet. The tallest buildings in New

10.2   Two triangles showing the
relation between R, d, and h for the
case of objects beyond the apparent
horizon

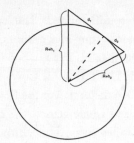

Haven are about twice this height and should be visible under ideal conditions and greater optical magnification. After all, the tops of the Pocono Mountains in eastern Pennsylvania lay at an even greater distance, but their summits at about 2,000 feet rendered them easily visible.

Just whence came the first realization that our world is round? We can't see or prove its rondure from any single observation, since a curvature or a dip in the horizon may be the result of local topography. But in classical Greece by the fifth or sixth century before the common era, Pythagoras and his followers appeared to know that we live on a globe. The first unequivocal statement and proof in print are from Aristotle (384–322 B.C.) writing in Athens. He realized that during every lunar eclipse the edge of the shadow of our Earth on the Moon's surface is always the arc of a circle, and the only solid shape that appears circular from every aspect is a sphere. This is still considered to be a valid proof of a round world. Since the time of Pythagoras, the educated world has held quite firmly that our planet is a globe.

Other proofs include the appearance of a ship leaving port to be sinking into the water as it sails away, regardless of its direction or the location of the port. It is not sinking, but rather sailing over the edge of the horizon. Similarly, the horizon lies below the horizontal plane when seen from a height. This is called the dip of the horizon and becomes evident at modest heights. The visible horizon is seen to be lower than the nominal horizon by an angle equal to the angle at the top of the line R + h.

To be certain that we live on a ball, we would need to replicate this perspective in many other spots all over the world to be sure that the rate of curvature is constant. Seen from a great altitude, however, the Earth is very round indeed. The curvature, now such a part of our planet's imagery from space shots, was first seen directly at one glance on November 11, 1935, when Captains Albert Stevens and Orvil Anderson of the U.S. Army Air Corps ascended in a balloon to a record 72,395 feet (22,066 meters) over the Black Hills of South Dakota. At that height they saw, at a single glance, the curved horizon of a sphere for the first time, 330 miles away. This altitude record for a manned ascent of any kind stood for years, until well after the Second World War. It had been brought to my attention by Jean Piccard, a friend and neighbor, whom I revered as a youngster. He and his twin brother, Auguste, pioneered open-gondola balloon flights before the war.

Yes, certainly Goethe could have seen the San Marco campanile from Padua on a clear day, and Galileo may have seen Padua from its balcony, when he demonstrated his telescope to the Doge there in 1609. The campanile they saw is not the one we see there today. That one just collapsed and crumbled into rubble in 1902 like a house of cards, but the city fathers of Venice of the time replaced it with an exact duplicate on the same spot, also 98.6 meters (about 324 feet) in height. This is the one we see today, usually through much more smog and air pollution than Galileo and Goethe had to contend with. How nice it is to view the distant terrain through the natural haze alone, lending through the effect of chiaroscuro a dusty grayness to the more distant mountain slopes without our common air pollution blanking them out altogether.

❋

IN THE DISCUSSION above, our needs were served by the simple model of the world represented as a sphere. Not one observation there would be perceptibly different on a slightly oblate Earth than it is on a sphere of the same size. This is how matters stood from the time of Aristotle until 1687, when Sir Isaac Newton published his work on gravitation. In fact only about two centuries after Aristotle's time, some of his successors living at Alexandria demonstrated that the Earth is round and even measured its size. Eratosthenes was one of the brilliant astronomers practicing his trade in the Egyptian city. On a trip south up the Nile

River to Syene, now known as Aswan, he noticed that the Sun at noon when farthest north in late June illuminated the bottom of a well. For that to happen, the Sun must be right overhead, at the zenith.

Back home in Alexandria, he found that the Sun could not do that since it never missed the zenith by less than 7°, passing always to its south. He knew or estimated that Aswan lay to the south of Alexandria at a distance of 5,000 stades (a unit roughly the diameter of a stadium then and now). At 10 stades to our mile, a 7° curvature amounts to 500 miles. Then a straightforward ratio applies, for 7° to equal 500 miles, the full 360° of the circumference of the globe must be around 25,200 miles, just 1 percent above its modern value.

It would be marvelous if that were the end of the tale, but it is not. We do not know exactly how many stades equal 1 mile, maybe 10, maybe 11 or 12. In the latter cases, Eratosthenes found an Earth maybe 10 to 20 percent smaller than we know it to be. Christopher Columbus seized upon the smallest of the allowable sizes for the globe when he applied for a travel grant from Ferdinand and Isabella. Since the size of Eurasia was known, he could present the best case for a sea route west to Cathay by wrapping the continent around a smaller sphere, much as wrapping part of a grapefruit rind around an orange covers more of the smaller orange than it did the larger grapefruit. A greater span of longitude taken up by Eurasia meant a smaller sea distance westward from Spain to China.

But here we need a more accurate model for the shape of the Earth, which is not quite round. This newer model derived directly from Newton's work. He realized that the existence of the precession, the very slow gyroscopic spinning motion of our axis, was due to the gravitational attraction of the Sun and the Moon upon a previously unsuspected equatorial bulge of the Earth. Our diurnal rotation gives rise to a centrifugal force that causes the world to bulge at its equator, as it must for the Moon and Sun to impose this precessional motion on it. As in the case of the tides, the Moon's effect is about twice that of the Sun, and since no other world is either very close like the Moon, or very large like the Sun, the effect of the planets is negligible in all but very precise work.

The precession was discovered by Hipparchus, a successor to Eratosthenes, in the second century B.C. Neither he nor anyone before Newton knew the reason for it, because the physics just wasn't well understood. After Newton called attention to the bulge, it was confirmed by the measurement of the actual length of 1° latitude. Were the Earth truly spherical, it would be the same, about 69 miles, at every latitude. But, as can be seen below in Figure 10.3, this is not so. The length was found to be about 68.7 miles near the equator and 69.4 miles near the Poles. At either Pole, the rate of curvature is seen to be equivalent to a larger true sphere; at the equator, a smaller one. We live on a slightly flattened spheroid, a better model of our world than the sphere. What are the implications of this model?

In Ecuador, just 1½° latitude south of the equator, stands that country's highest mountain. This is Chimborazo, reaching 20,561 feet (6,267 meters) into the Andean sky. No other mountain so close to the equator soars to such an altitude above sea level.

What is unique about this mountain? Mount Everest is widely known to be the highest peak in the world; in fact, much of the great Himalayan massif extends higher than any summit in the Andean cordillera, or anywhere else on the planet.

The answer lies in the realization that the Earth is not round—not quite. It is an oblate spheroid, a shape defined as having one of its three mutually orthogonal (perpendicular) axes shorter than the other two (an object, such as a football, with one axis longer than the other two is known as a prolate spheroid, and one with all three axes of different lengths is an ellipsoid). Thus along any diameter in the plane of the equator, the world measures 7,926.41 miles, whereas from one Pole to the other, it is smaller, being 7,899.83 miles in size. This forms a ratio of flattening of only 1 part in 298, an amount much too small to detect in images of the Earth from space, or to concern globe makers, who make

10.3   Curvature of the Earth at the North Pole and at the equator. The Earth, being oblate, has a curvature of a smaller sphere at the equator and a larger sphere at the Poles. Its oblateness is greatly exaggerated to reveal the difference.

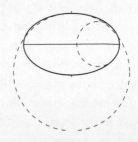

spherical globes. On a basketball, this same ratio would render the equatorial diameter larger than the polar by a mere one thirtieth of an inch, a little less than a single millimeter. Michael Jordan forced a basketball out of round by that much with every free throw.

Some of the other planets are flattened much more obviously. Jupiter, seen through any small telescope, reveals its equatorial bulge at once, being one part in fifteen larger across its middle than along its polar diameter. That huge planet rotates with a period of only 10 hours with a size eleven times the Earth's and thirteen hundred times its volume; no wonder that centrifugal force molds this great bulk well out of round. The champion is Saturn, whose equatorial size is one tenth larger than its polar axis. At nine times the diameter of our planet, it is smaller than Jupiter, and its day is just a bit longer, at 10½ hours. But Saturn is much less dense than Jupiter. Indeed, Saturn is so light that it would float on an ocean large enough to contain it.

Although a seemingly trivial amount, the Earth's flattening is much larger than its topographical variation. Thus, the summit of Mount Everest, at 29,028 feet (8,848 meters) above sea level and situated at a latitude of 28° north of the equator, is well short of being the farthest point from the center. On our basketball model, this tallest mountain soars but one hundredth of an inch above the rest of the terrain. It may be worth noting that a new measure of the altitude of the summit of this highest of all mountains has added 7 feet to its height; it is now officially 29,035 feet above sea level.

A point at sea level on the equator is 13.3 miles farther from the center than is a sea level point at either pole, 13.3 being the difference in miles between the equatorial and the polar radii. The summit of Mount Everest is equal to 3,949.92 (the polar diameter) plus 9.16 miles plus its height above sea level in miles, 5.50, for a total of 3,964.58 miles, whereas the same quantity for Chimborazo turns out to be 3,966.89 miles, or 2.31 miles (about 12,200 feet) farther from the center of the Earth.

Thus, the peak of the Ecuadorean mountain is the farthest point on the surface of the world from its center. The closest point must be at one of the poles. The South Pole stands atop the Antarctic plateau at about 9,000 feet above sea level. The North Pole, on the other hand, lies amid the Arctic Ocean right at sea level. Thus the North Pole is the closest point, 3,949.92 miles distant from the center.

*

WOULD WE FIND a difference in our weight if we moved between Chimborazo and the North Pole? The answer is yes, and Newton's laws tell us about what that difference should be. Newton realized that the force of gravity was directly proportional to the product between any two masses divided by the square of the distance between their centers. Every object attracts every other in the universe according to this law, making it a true universal law.

This is illustrated by the weight of an object. We define the weight as its mass as attracted by the Earth when at its surface. As Newton proved, its attraction is the same as it would be if all of its mass were compacted to the point of its center. Our mass is set equivalent to our weight at the surface by definition. On the Moon, one's mass would be the very same but one's weight would be much less. The Moon's radius is only about 1,000 miles, a quarter of ours, but its mass is only $\frac{1}{81}$ as much as the Earth's. That works out to one sixth the gravitational attraction at its surface. A 180-pound person would weigh only 30 pounds on the Moon. He would be as fat as ever, but his weight loss would exceed any diet plan that might be conjured up for us Americans to try. He would even weigh less than the concentration-camp-thin supermodels so fashionable these days.

Think of the sports possibilities there in a domed lunar stadium filled with air. With no cumbersome space suits to carry around, John Elway or Dan Marino could hurl a forward pass over 300 yards, three normal football fields. The receiver could leap 10 or 15 feet upward to catch it, as could the defender. The record field goal would be near one fifth of a mile! Mark McGwire's and Sammy Sosa's longest home runs would sail more than half a mile into an outfield that would require an acreage larger than the combined sizes of the fields of all of the present major league baseball stadiums in North America. On a lunar golf course a par 5 hole might be 2 miles long, and a bench press of 1,500 pounds would not be particularly impressive in weight lifting. Clearly the first lunar Olympics would produce a spectacular set of track and field records.

Now imagine a tower on Earth 4,000 miles in height. At the top of the tower a person would be twice as far from its center as at its surface. How much would he weigh at the top? Since the distance has been doubled, the gravitation would be only one fourth as much, and he would weigh 45 pounds there. This knowledge can be used to derive the range in weight between Chimborazo and the North Pole. The ratio of their distances stands as 1.0043 to one. Thus the weight difference is 0.99146 to one and 180 pounds at the Pole amounts to 178.46 at the mountain peak. The average adult weighs 1 to 1½ pounds less at that summit, due to the greater distance from the center.

But Newton was also aware that the rotation gave a second impetus in the form of centrifugal force, which is nothing more than the tendency for an object to continue in a straight line. Other factors of a more geophysical nature also enter in so that a definitive quantity is not easy to determine. We can assume about 0.5 percent weight loss for objects moving from Pole to equator plus an additional weight loss of about 0.2 percent for climbing to the summit of Mount Chimborazo, excluding weight loss from loss in fluids during the climb.

The maximum range in weight over the Earth's surface is perhaps about 1 pound for an average adult, providing one travels to two very remote and almost inaccessible points on it, and providing the right kind of scale is brought along. A balance, or a scale in a doctor's office with weights sliding along bars, would show no difference at any site, or atop the tower or on the Moon, because the weights would also vary up and down in the same ratio. But a bathroom scale registers the weight against the tension of springs, and this would be constant everywhere. This type of scale would reveal the differences mentioned above.

11

*

Brahms, Coincidence, and the Star of Bethlehem

Correlation is not causation.
Paul and Anne Ehrlich,
Betrayal of Science and Reason

Johannes Brahms was born in the city of Hamburg on May 7, 1833, and died in Vienna on April 3, 1897. He was one of the handful of composers among the favorites of both my father and myself. It might strike one as odd and even significant, then, that my father was born just 75 days before Brahms died, and that I was born exactly the same number of days before the centennial of his birth.

To what degree can we simply put this down to coincidence, or might we be justified in ascribing it to the world of the paranormal? An answer to this question could be most sensibly obtained from the frequency of such patterns within the entire spectrum of human experience. This is not easy to calculate in a solid, quantitative way, but in any event, the supernatural is sought much more frequently than it can be justified. How often among the members of one's family might such an unlikely set of events arise— there may be others that are equally singular, of which I am unaware. This one coincidence is counterbalanced by the many that might have happened, but did not. It is not the case for Beethoven, Mozart, Schubert, or any other composer, nor, to my knowledge, for any other person.

Often we hear that someone is mystified and delighted because, for example, his or her birthday falls on the eleventh of the month, a sister's on the fourteenth, her roommate's on the fifteenth, and a nephew's on the seventeenth. Isn't that just remarkable? No, probably not, since among anyone's family and close friends a grouping of this sort is pretty likely to show up somewhere.

Two such happenstances, well known in American history, illustrate near-commonplace episodes often taken as something more. Both deal with our presidents. The only two presidents among the signers of the Declaration of Independence, John Adams and Thomas Jefferson, died just hours apart on the day of its fiftieth anniversary, July 4, 1826. But what if this had not happened, and instead, the two Nobel Laureates among our chief executives, Theodore Roosevelt and Woodrow Wilson, had both died on January 15, 1929, the day that Martin Luther King Jr., another winner of the Nobel Peace Prize, was born? Would we not saturate that event with equal or even greater wonder?

Perhaps even more remarkable and better known among the presidents is the array, considered a reliable jinx by some, of the deaths in office of just those elected in years divisible by 20, or at least until Ronald Reagan spoiled the rule in 1989, when he left office alive. It is often conveniently forgotten that the first two presidents to qualify, Jefferson and James Monroe, elected in 1800 and 1820, also finished their terms, and that Zachary Taylor died in office after being elected in the year 1848.

Coincidences such as these, made up as they are of a string of single events, are sometimes taken as meaningful, as being bound together by a causal and possibly paranormal relationship. A further continuation of nearly adjacent birthdays or an extension of presidential deaths in office to just those elected in 2000, 2020, and beyond would be needed before an extraordinary explanation can even begin to be realistic. Just how many is problematical. There could come a time when the 20-year presidential jinx should be taken seriously, but that would require perhaps another consecutive or near-consecutive string of deaths in office every 20 years for at least a century or two, perhaps longer. Only then might a reason emerge for the so-called jinx.

Repeated events can have an exasperating way of lining up to suggest an apparent cyclic pattern, only to break down upon extension. Just when it seems that one phenomenon is thought to depend upon another, an exception comes along to spoil it. Just as soon as the conference of the winning team in the Super Bowl or the league to which the winner of the World Series adheres appears to foretell the Dow Jones Index or the outcome of an election, it breaks down.

In astronomy, observers note coincident events with great frequency. How often do the two brightest nocturnal celestial objects, the Moon and Venus, get close together in the sky? If we limit the discussion to the evening sky, whenever Venus is seen and often called the evening star, the approach happens once every lunation, a lunation being the period between two occurrences of the Moon at a particular phase (between one new moon and the next, for example). This period averages 29½ days. It is 2⅙ days longer than the true orbital period of 27⅓ days of the Moon about the Earth because between one new or full moon and the next, the Earth and the Moon orbit about one twelfth of the way around the Sun, and the lineup must take that motion into account for the Moon to "catch up" to the Sun in the sky. But the phase of the Moon is by far its most notable feature; thus, we mark months by the longer period of time.

Being closer to the Sun than the Earth, Venus appears to swing from one side of the Sun to the other about once every 19 months. It spends nearly half of the time appearing as the evening star after sunset and nearly half as a morning star rising before dawn. For a small time it is too close to the Sun in the sky to be seen at all. Since Venus is never seen far from the Sun, the Moon when passing it in the sky is always seen in the crescent phase. The crescent Moon and Venus seem to hang there together in the twilight, both so bright that they are visible even at sunset, and they dominate the night sky soon afterward. Over the interval of 19 months, then, the two are close 1 night of every 29 or 30, for about 8 or 9 months in a row, and then are absent in the evening sky for 10 or 11 months. It is by no means a rare sight but whenever these two line up, the whole world takes notice—it is spectacular but not rare.

Near the rare end of the scale of planetary configurations are the close conjunctions of Mars, Jupiter, and Saturn, the three bright superior planets (meaning farther from the Sun than we are; Mercury and Venus are inferior in that respect). These occur but once in about 800 years. Johannes Kepler, the great astronomer who established the basic laws of planetary motion, noted such an event in 1603 and determined that the previous clustering took place about 800 when Charlemagne was crowned as the ruler of much of western Europe. The conjunction prior to that one took place about 4 B.C. at nearly the correct time of the birth of Christ, and may have served as the star of Bethlehem. These conjunctions were seen as powerful omens by astrologers, and surely touched off much speculation about earthly events. As celestial "signs" or "happenings," these events could have been described as stars, especially after the translations the story has passed through from one language to another. The only other candidates for the star seen by the wise men appear to be an exploding star or supernova, or a bright comet.

Jupiter and Saturn approach each other about once every 20 years, and Kepler showed that every fortieth of these meetings Mars is also on hand. In 20 years, the Moon and Venus will meet in the evening sky around a hundred times, but these occurrences are much more often seen as spectacles than other conjunctions because both objects are so bright. Perception is a function of visibility as much as it is of infrequency. Even now, as this is being written, a grouping of the Moon and most of the planets is occurring, and if the blinding Sun were not in the midst, they would mount a spectacle in the night sky.

There are, to be sure, other equally plausible explanations for the star of Bethlehem. It may have been a supernova, a massive star that blows up, briefly rivaling its whole galaxy in total brilliance. Supernovas are rare; the last two in the Milky Way were seen in 1572 and 1604. At least the first and maybe both could be seen in the daytime sky. A supernova appearing in 1006 holds the record for the brightest starlike object seen in recorded history. At a magnitude near –9, it was almost as bright as the first-quarter Moon! Another came along in 1054; also a daytime object, it was first noticed on July 4 of that year. Its remains are splattered all over its corner of space and are known to us as the Crab Nebula. Supernova explosions are limited to very massive stars; the Sun and its nearby neighbors will die less drastically. Of all bright stars, Betelgeuse, the bright red supergiant star gleaming in Orion, has probably the best chance of any familiar object of mounting such a display. Should it do so, it might well rival or surpass the Full Moon in brightness for a few days or weeks. This is fairly likely to happen sometime in the next 10,000 or 100,000 years. But it might just possibly blow up tomorrow night. Something like a supernova, or even a nearby nova, in which a star blows off a part of its outer layers, would certainly command the attention of the wise men and everyone else. The event could also have been marked by a bright comet, but this is unlikely because very few comets can ever become bright enough to be visible in the daytime.

Brightness predominates in notice and perception of objects in the night sky. As we mentioned earlier, we can see at least two of the objects of 0 magnitude at any time from any place on Earth. But among brighter objects we find only the two brightest stars in the night sky, Sirius (available only during our winter) and Canopus (invisible north of the latitude of Atlanta or Los Angeles). Venus and Jupiter are always much brighter than this, but Mercury is seen only in the twilight and Mars is of −1 magnitude or brighter only for a few months every other year. Frequently none of these celestial standouts can be found, and only rarely are more than two visible at the same time.

It may not be surprising that of the hundreds of requests I have received for the identification of that bright star up there, every one has come down to just five bodies. Venus at −4 magnitude accounts for most of them, Jupiter at −2 is next, and only Mars, Canopus, and Sirius have also ever provoked this question. It seems likely that the 0-magnitude stars are seen by people as commonplace and only the brighter five, and particularly Venus, are appreciated as extraordinary. Moreover, I have spent only a small part of my life in latitudes from which Canopus clears the horizon; yet I have been queried about its identification, while I have never been about the 0-magnitude bunch. In 1975, when a nova burst forth in the northern constellation Cygnus the swan, few noticed it, because it became of the second magnitude at its brightest, rivaling the Pole Star, but not the brightest of its neighbors. Perception is here much more a function of brilliance than of rarity.

12

*

Transits and Other Syzygies

Our solar system is a crowded neighborhood, particularly here in its inner regions. We have one star, eight or nine major planets (Pluto is so tiny that its classification as a planet is more a question of semantics than astrophysics), about sixty satellites, and four sets of rings around each of the four major planets, perhaps fifty thousand asteroids, minor planets that orbit the Sun mostly between Mars and Jupiter, and a few that lie out near Pluto made mostly of snow and ice, some interplanetary dust lying around, and zillions of comets and small meteors. Most of the comets hang out far beyond Pluto in the Oort cloud, named after its discoverer, Professor Jan H. Oort.

Occasionally three among this multitude of objects form a syzygy, a close alignment such that the object in the middle passes in front of one as seen from the other of the two objects at either end of the line. The best known syzygies are the eclipses of the Sun and the Moon. In the case of a solar eclipse, the Moon covers all or part of the Sun as it passes in front of the solar disk. A lunar eclipse is not strictly an eclipse at all since the Moon remains visible, but the Earth cuts off most of its sunlight, leaving the lunar surface shining faintly by a little light that is refracted by our atmosphere. Those on the Moon would experience a true eclipse since the Earth would be seen to cover the Sun.

A number of other terms cover other forms of syzygies; the word *occultation* is used if the eclipsing object appears much larger than the one covered. A *transit* is taken to mean just the reverse, where a small dot is seen against the flank of a large planet or, most likely, the Sun.

The Moon is seen to occult planets and stars fairly frequently (so does the Sun, but who can see them?). Planets, being much smaller than either in angular size, do so only infrequently. Exceptions to this rarity are planets that pass in front of their own satellites. Jupiter does the best job of this; during every orbit at least three, and sometimes all four, of the large ones pass behind Jupiter's bulk. And they also pass in front, but here the transits are seen as tiny dark shadows cast onto the great flank of the giant planet as its moon rushes by.

The transits of by far the most interest to astronomers are those of Mercury and Venus, the two planets closer to the Sun than the Earth is, and thus are able to pass between the two. Just as the Moon's orbit is inclined by about 5° from the ecliptic—the Sun's path across the sky—so is the orbit of each planet also tilted slightly to the plane of the ecliptic. All orbits are seen as great circles on the celestial sphere, and so each must cross every other at two points, opposite each other in the sky, called nodes. Transits of either planet across the face of the Sun can take place only when both Sun and planet are near a node. As it happens, Mercury can transit the Sun only in May or November, and Venus only in June and December, because only at those times are they near the nodes, the points where the two orbits intersect.

Mercury is a small planet, appearing only about 10 seconds of arc in diameter when in transit. Venus is a good deal larger and is closer to us; its disk is about 1 minute (60 seconds) of arc when it appears in transit. Neither planet is visible to the naked eye in the course of these events. In order to detect a transit, we had to await two developments. The first was the telescope, invented in 1608, and improved and used to explore the heavens by Galileo in the following year. That same year, Kepler published the first two laws of planetary motion, with the third and final law coming ten years later.

It was only after Kepler that anyone had the means to predict where planets would be seen in future years. Only then could a transit be predicted. It comes as no surprise, then, that the very first transit to be observed was in 1631, of Mercury, despite the fact that four others had occurred since 1610 when telescopes were available. Kepler had made two startling predictions; he predicted that both planets would transit that year, Mercury on November 7 and Venus a month later, on December 6. The latter event would not be visible from Europe, but Mercury's crossing would, and a French astronomer, Pierre Gassendi, saw it from his apartment in Paris. He did as all must do; he projected the image of the Sun onto a screen behind the eyepiece, since no one could or ever should look at the Sun directly through any telescope. He saw the tiny black dot travel across the huge Sun and attempted to calculate Mercury's size.

A young English astronomer was the next to predict a transit. Jeremiah Horrocks, who died in 1641 at the age of twenty-three and was a great loss to astronomy, found that Kepler's calculations were not free of errors and also found that he had missed predicting a transit of Venus in 1639. Horrocks and a friend were the first to watch a Venus transit that year.

Transits of Mercury are infrequent; there were fourteen of them in the twentieth century and fourteen more will happen in the twenty-first century, always in early May or early November. The intervals in between them range from 3 to 13 years, with the next due on May 7, 2003.

Venus is more capricious in exposing her dark side to us, as befits a goddess of love and beauty. The fair planet almost but not quite duplicates her motions every 8 years. This means that transits come along in pairs separated by 8 years less 3 days, but the pairs are separated by more than a century. Since the coming of the telescope, they have taken place in 1631, 1639, 1761, 1769, 1874, and 1882—none in the twentieth century. But soon, and this is the primary raison d'être for this chapter, on June 8, 2004, and again on June 6, 2012, Venus will step again in front of the Sun. Other transits will follow, but not until the years 2117, 2125, 2247, 2255, 2360, and 2368.

In addition to a much larger black disk, Venus presents an entirely different face than Mercury. The smaller planet, like the Moon, has almost no atmosphere—it is just too small and too hot there so close to the boiling Sun, but Venus has a very thick one. Its atmosphere is ninety times as thick and dense at the surface as ours! Furthermore, being mostly composed of carbon dioxide, the atmosphere has experienced a "runaway" greenhouse effect. The temperature at the Venerian or Cytherian (adjectives for Venus) surface is some 870°F (500°C), hot enough to preclude any chance of carbon-based life, or likely any other kind, from getting started. This is much hotter than it would be if heated by the Sun alone; hence the lovely evening star, far from being a lush world of luxury and opulence, bears a close resemblance to Hell. This atmosphere permanently enshrouds the planet in clouds and, when seen before the Sun, exhibits itself as a bright ring around the disk, refracting light into the shadow behind it just upon ingress at the start of the transit, much as the Earth does the Moon during a lunar eclipse.

Astronomers are excited by transits, and understandably so. Edmond Halley in the late seventeenth century postulated that transits provide a means of determining the most important distance in the universe, the astronomical unit or mean distance between the Sun and the Earth. In Kepler's day, that number was thought to be about one third of the 93 million miles we know it to be. But Newton and Halley found ways of measuring it more precisely, if only indirectly. Their distance of 87 million miles was at least close to the correct value. Attempts to measure the change of position of the resplendent solar disk against a reference frame of stars would be as ludicrous then as they would be today. But once the relative scale of the planetary system was known, observations of other planets can be translated into the all-important astronomical unit.

From Venus or Mercury, there would be nothing to see. The Earth would simply be located more exactly opposite the Sun than it would be most other times when it passes in opposition to the Sun; there's no fun in watching that. But of course if anyone should establish residency on Venus, there are transits of Mercury to watch, fifteen of them in the next 33 years. But it is from Mars that the shows become frequent. It has three planets and one large satellite to keep track of; Mercury dances in front of the Sun every few years, Venus does so about seven times each century, and the Earth once or twice per century. And our Earth comes with a dividend, for the Moon would on most occasions flit along with the larger Earth across the Sun's face.

The last transit of the Earth as seen from Mars happened on May 11, 1984, and inspired Arthur C. Clarke to write a short story about it well before the event itself; the story, "Transit of Earth," appeared in *Playboy* magazine in its January 1971 issue. In his tale, a single astronaut remains alive after his spaceship crashes onto the Martian surface and he spends his last hours broadcasting what he sees to anyone back home on Earth who might happen to be listening. All of the facts mentioned in the story were exactly known in 1971 (13 years before the event), having been derived by Jean Meeus, a Belgian astronomer. No one, of course, witnessed that event, but by the time the next Earth transit comes along, on November 10, 2084, just 100 years after this last one, there may well be someone on Mars to witness it.

Transits of the four terrestrial inner planets would be difficult to view from any of the outer four major planets. Jupiter and Saturn, and especially Uranus and Neptune, are just too far out there to have reason to take much interest in these occurrences. Might a transit of Jupiter be observable from Saturn? Of course it would—sometime. This raises an issue of some interest. Whenever Jupiter crosses the solar disk, it cuts off at least 1 percent of the Sun's light, being one tenth its diameter (more from Saturn because Jupiter is much closer and would appear larger). But even at nearly the same relative distance, as is the case for other stars, this 1 percent of light corresponds to 0.01 magnitude. That is a diminution we can see with a photoelectric photometer, which surpasses this level of accuracy. Therefore astronomers have launched a program of observation in which stars similar to the Sun are monitored in order to detect any loss of light, implying a transit of a major planet. The planet's existence will be confirmed should such an event occur. Jupiter-sized planets have already been proven by indirect means to orbit a few dozen nearby stars, but here a direct observation could confirm other planets as well. ·

13

*

The Clockwork Sky

Difficulty of empathy, of genuinely entering with the mental and emotional values of the Middle Ages, is the final obstacle. The main barrier is, I believe, the Christian religion as it then was: the matrix and law of medieval life, omnipresent, indeed compulsory. The insistent principle that the life of the spirit and of the afterworld was superior to the here and now, to material life on earth, is one that the modern world does not share, no matter how devout some present-day Christians may be. The rupture of this principle and its replacement by belief in the worth of the individual and of an active life not necessarily focused on God is, in fact, what created the modern world and ended the Middle Ages.

Barbara W. Tuchman, *A Distant Mirror*

Every period of time by which we mark off on the calendar our daily, monthly, and yearly affairs is not a true measure of anything! The Earth does not spin once upon its axis in 24 hours, the Moon does not orbit the Earth in 30 or even 29½ days, and the Earth does not circle the Sun in 365¼ days. This is not all a sham, something astronomers cooked up years ago, but a natural consequence of our desire to keep track of time by the Sun, a need that predates civilization.

We can illustrate how this happens by imagining a thought experiment. Suppose the Sun stands high in the south, and as a consequence it is noon. Now suppose we can turn down the wattage of the Sun until it is only as bright as the Full Moon, so that the brighter stars are visible. Now imagine a rather bright star just above this dimmed-down Sun, both due south in the sky. Many planetaria can do this to the Sun they project, but it is best that we cannot so meddle with the real one. Now we wait 24 hours marked off by a very accurate clock. Once again, weather permitting, we see the Sun due south. But the star is not there; it is about 1°, about twice the Sun's apparent diameter, to the west of the Sun, to the right of it as viewed from the Northern Hemisphere. This came about because in the course of a day, we orbit the Sun about ⅟₃₆₅ of the year, and with 360° around the sky, the Sun appears to move eastward about 1° each day. If we could stop the rotation of the world, as Joshua is alleged to have done, the Sun would slowly move 1° eastward every 24 hours. In about 3 months it would set in the east, and 6 months after that it would rise in the west. After an entire year, it would get back to due south again.

The Earth spins through all 360° in the sky in 24 hours, or 15° per hour. One degree takes only 4 minutes to move past any fixed point in the sky, such as marked by our bright star, so the star would reach the due south position in the sky 4 minutes earlier each 24-hour day. After one day, it would pass due south in 23 hours and 56 minutes, and then 4 more minutes must elapse for the Sun to reach that direction (see Figure 13.1). What, then, is the *true* period of rotation? It is 23 hours and 56 minutes. The star pulls ahead of the Sun by 1° (4 minutes) every day, and like a slightly faster car on a racetrack finally laps the Sun once per year. In that year the Sun has gone 365¼ times around the track (sky) while the star covered 366¼ laps. We want to conduct our affairs by the Sun and we don't care what any other star is doing. We want to rise when the Sun does or soon thereafter, and go to bed when it sets, or a set number of hours afterward, and lunchtime should come at noon when the Sun is high in the sky. Especially do farmers, perhaps more than anyone else, need to work and live by the Sun and only the Sun. Even a tyrant like Stalin could not vary this regimen, as he once tried to do when he was alleged to have decreed that all of the Soviet Union would adhere to the time zone of Moscow.

Our true day is in reality only 23 hours and 56 minutes long but if we made that our working day, we would keep time by the stars. This kind

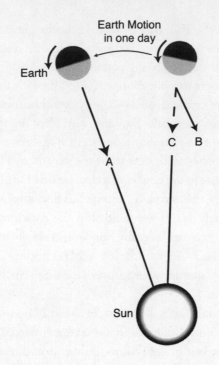

13.1 The difference between the sidereal and solar day is due to the orbital motion of the Earth about the Sun.

of time is called sidereal time and is used by astronomers when they observe stars and not the Sun. If we ran our legal and social affairs by the stars, we should find that waking time, in what would pass for morning, would occur when the Sun is high in the south at one time of the year and setting at another. Similarly, dinnertime might see the Sun rising, setting, and down altogether at different seasons. No civilization ever has or would countenance such confusion, and farmers would go nonlinear when attempting to deduce the proper time to plant or to harvest.

The same thing happens in the motions of the Moon. By far the most obvious feature about the Moon is its set of phases. As remarked earlier, the time it takes the Moon to repeat a phase, to go from one new Moon to the next, for instance, is just about 29½ 24-hour days. But if this new Moon occurs when it and the Sun line up next to our star, the next new Moon will find the star about 30° to the west of it and the Sun. Again in the intervening period, the Earth, carrying the Moon along with it, has orbited about one twelfth of the way around the Sun (see Figure 13.2). It had taken only 27⅓ days for the Moon to again reach the star, but the Sun was no longer there; it was off to the east, and 2 more days were needed for the Moon to catch up. So analogously, the sidereal month (the interval between two successive passes of the Moon past the star) is but 27⅓ days in length. But we use the 29½-day lunation to mark the month, and about twelve of them make a year, or close enough with a little juggling. Many mideastern calendars are based on twelve of these intervals, amounting to 354 days, about 11 days shorter than the true year. So after a few 354-day years, a thirteenth month would be inserted, making a 384-day year. Practice showed that

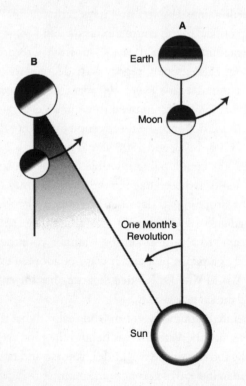

13.2   The difference between the sidereal and synodic month is also due to the orbital motion of the Earth about the Sun.

seven of these extra months inserted every 19 years balanced things out rather nicely. The Jewish calendar and others in that region are formed along these lines and are known as lunisolar calendars, as opposed to the solar calendars such as our Julian and Gregorian schemes.

Nor is the year exempt from this hybrid kind of motion. If the Sun and our convenient star are in line right now, about 365¼ days will pass until the next lineup. The Julian calendar used just this period as the base for its calendar. In 46 B.C., Julius Caesar decreed its use throughout the empire to avert the many different calendars then in use, some of which were corrupt. This length of time is not quite correct. In the eighth century, the Venerable Bede, for one, knew that this period was too long by about 11 minutes and 48 seconds. Not much in 1 year, but over the centuries the difference accumulated and holidays were observed when the Sun was, by Bede's time, about 3 or 4 days early. There went the proper date for Easter, critical to the church in his day, and something had to be done. This "something" took 850 years—possibly a record for procrastination. Finally in 1582, Pope Gregory XIII decreed that since the slippage was very close to 3 days every 400 years, he ordered the extra leap year day, February 29, to be removed three times every 4 centuries. The years in which this is done are the three century years not divisible by 400. Thus the years 1700, 1800, and 1900 would not have a 29-day February, but 1600 and 2000 would, being divisible by 400. But since Easter and other movable feasts had been fixed in the fourth century, the pope had to realign them by dropping 10 days out altogether; October 4, 1582, was followed immediately by October 15. The peasants were confused by all this but they came to accept it—in the Catholic countries. Elsewhere it was denounced as a papist plot and it was not adopted everywhere until after the First World War. Only since that time has the entire world celebrated one and the same calendar.

But long before that time even England realized that the alignment of the solar system, not the Vatican, was behind the plot, so in 1752 it went along; gradually the rest of the world did likewise and now we have only one legal calendar. Instead of adding some minutes to New Year's Eve, we add a day every fourth year, and then drop that thrice in four centuries. Things are not quite square even now with our Gregorian calendar. We are still off by about a day every 3,000 years! So sometime after A.D. 3000, the United Nations or whatever has replaced it will have to make another adjustment in order to fine-tune the proper date of Easter.

And so it goes. Even the period of precession has to go by two definitions. Its true period is just about 25,800 years. But in that time some yet-longer periods will come along and mess things up again; for climatological purposes, the period is adjusted down to about 21,000 years. The true period of anything is called sidereal. We have a sidereal day, a sidereal month, and a sidereal year, but except for a few astronomers and other eccentrics, we use none of them.

Other time periods delineate other intervals. We have 24 hours to the day, 60 minutes to the hour, and 60 seconds to the minute. These are conveniences, but the week has another origin. The 7-day week, not universal, comes not from some obscure celestial motion, but it does come from the sky. Back in the days when the planets were considered to be gods or manifestations of gods, there were seven of them. Five are the five bright naked-eye planets: Mercury, Venus, Mars, Jupiter, and Saturn. But the Sun and the Moon were also considered planets because they, like the other five, are soon seen to move across the sky and thought at the time to also circle the Earth. All of the stars appear fixed; they move mighty slowly so that in one lifetime they cannot be seen to move without telescopes or other very precise instruments. But, with each of the moving seven being considered a god, each got a day to rule and the Sun, being much the brightest and most important, got the Sabbath in most cultures.

In classical times, Mount Olympus, the highest peak in Greece, was considered the abode of the gods, and its dramatis personae in residence included among others the names of those fortunate to have a planet named in their honor. The Roman names are the ones we use; in the customary order outward from the Sun they are Mercury, Venus, Earth, Mars, Jupiter, Saturn, Uranus, Neptune, and Pluto. The corresponding names in Greek are Hermes, Aphrodite, Terra, Ares, Zeus, Chronos, Uranus, Poseidon, and Hades. Only Uranus, the god of the sky, retains his name in the Roman transformation.

As the planets, including the Sun and the Moon, were found to move in predictable ways, it has been uncertain as to whether a particular society would view them as gods or as celestial manifestations of gods. Whichever was the case, people everywhere began to perceive influences of each planet on one's daily affairs. With the predilection so many have for a celestial mystique, astrology was a natural consequence. Astrology has flourished in the Middle East and on the Indian subcontinent since classical times or earlier.

A founding father, or at least a codifier, of astrology as we know it was Ptolemy, the second-century Alexandrian astronomer who wrote a book known as the *Tetrabiblios,* in which he discusses astrology, as well as the *Almagest,* a book that does the same for astronomy. With the rise of the Christian church, astrology flourished despite coolness toward it on the part of the Vatican. The early church looked with disdain on astrology because it tended to make gods of the planets, a polytheism that went against one of Christianity's basic tenets. Treating the planets as if they were gods or even manifestations of gods was bad enough, but the early church also held that free will governs the human soul and astrology favors a deterministic outlook in its predictions based only on the positions of the planets in the sky. The rise of charlatanism was a third feature in the disapproval of astrologers by the clergy. Nevertheless, the pseudoscience gained in favor throughout the Middle Ages, and customarily its practitioners included most astronomers up until the time of Galileo, and particularly Kepler.

It was Johannes Kepler, more than anyone else, who brought about a permanent severance between astrology and astronomy with his laws of planetary motion. By following laws that could predict the planetary positions in the future, there came a schism between the two that only deepened with the discoveries of Isaac Newton. Since his time a mutual contempt, even an enmity, has arisen to make of astrology a troublesome pseudoscience. With Newton, it became the mass of a body and its distance from the Earth that governed its amount of influence on our mundane world, not its position or direction in the sky or the sign or constellation in which the Sun happened to appear at the moment of one's birth. The divinity of the other planets never again became a concern for the science of astronomy.

14

✳

Battle in the Sky

One summer evening when I was perhaps ten or twelve, my grand-father and I sat together on the porch of his home in Wisconsin and he chanced to reminisce about his own childhood in Solvesborg, Sweden, where he was born. Solvesborg is on the shore of the Baltic Sea, and Grandpa loved to watch the ships coming and going on their way to Copenhagen or Saint Petersburg.

I recall vividly his telling me of the mirages common to southern Sweden and nearby Denmark, one of which he described in great detail. He saw above the waters of the Baltic a grand image of two armies fighting a pitched battle with the weapons of his day: cannon, mounted cavalry, and foot soldiers having at each other amid smoke and chaos. His memory of the specter gleamed, perhaps in Technicolor for all I know, and his dates for the vision were very explicit. He was puzzled because he knew of no war in Europe or anywhere at that time. Born in 1862, at the time of the American Civil War, Grandpa came to America in his twenties and was an old man when I was born. Even in his last years he was not given to confusion between real and unreal, and his wits stayed with him until he died in his mid-eighties. He had a lifelong interest in things historical and geographical, natural and man-made. One of his favorite books is still with me, *The Complete Book of Marvels,* by an adventurer named

Richard Halliburton who disappeared at a young age, like Amelia Earhart, while crossing the Pacific Ocean (on a raft in his case).

An opportunity to inquire about Grandpa's grand vision came much later in my life, after he was gone. My research in astronomy centered on the distribution of stars in space and the properties of the Milky Way to which they all belong. A number of Swedish astronomers work in this same field, at Lund in the south and at Stockholm and Uppsala farther to the north, and many became friends. Lund is only about an hour west of Solvesborg; there I raised the matter of Grandpa's spectacle and asked them their opinion of it. They confirmed what I surmised. Although mirages are not uncommon along the northern coastline of the Baltic Sea, they are all displacements of real objects. It couldn't have been any kind of atmospheric phenomenon; only in the world of the imagination could such an event take place.

As earlier noted, atmospheric refraction tends to lift images so that they appear higher in the sky than they really are. It is a kind of mirage, but most mirages, real mirages, appear in and because of the air around us. They are of two kinds. The most commonly seen are probably the puddles of water that appear to be lying on the road ahead, only to vanish as we drive closer to them. This is an example of an inferior mirage, one that arises from the heated air above a hot flat surface. The flat surface of a roadway, seen from a considerable distance, appears to reflect the bright and clear objects in the distance. What appears as water on the road is actually a reflection of clear sky afar. The mirage is seen when the roadway surface is very hot, heated by the Sun's radiation. The temperature of the air even a single centimeter above the surface, about half an inch, can be as much as 20°F cooler than the surface itself, and the air temperature continues to fall with height at a rate of a few degrees per centimeter.

This type of mirage is a consequence of the varying refractive property of heated air. The air nearest the surface is warmer and thus more rarefied, having a diminished refractive index. The light from just above is bent by passing through air of different temperatures. If a tree or other object can be delineated in the mirage it will appear inverted since rays entering from the hot, less-dense air just at the surface curve upward more than the rays passing a bit higher up.

A flat desert is another favorite place for inferior images or mirages. The very hot desert sands do the same as does the road surface, and desert travelers may be fooled into seeing water up ahead that isn't really there at all. Inferior mirages are captured on film much as they are seen by the eye. In the David Lean production of the film *Lawrence of Arabia,* among other desert movies, this was used to great effect.

Superior mirages are the result of very cold air, cooled by an icy or snowy surface underneath, with the coldest air just next to it and warmer air just above. The best known of these upward mirages is the fata morgana, so named from the legend that Fata Morgana, or fairy Morgan, half sister of King Arthur of medieval Camelot fame, was thought to live in a submerged castle that her magical powers could make rise into the air. This large mirage is observed when the air temperature increases with altitude above the surface slowly and then more rapidly. Sometimes real objects below the horizon due to the curvature of the Earth are projected upside down in the sky. Had a real battle been fought just over the horizon, Grandpa might just have seen a bit of it played upside down up there.

Rings around the Moon or the Sun come in two major varieties. One is a wide ring of about 23° in radius that, if seen complete, fills a fair portion of the sky. This is called a halo and happens at times when ice crystals are present aloft near the stratosphere. Ice crystals reflect light only at these specific angles and are associated with cirrus clouds, the thin feathery kind that are themselves made up of ice crystals and are too diaphanous to block light from the Sun or Moon.

The second kind of ring is the corona. Coronas are brightest right at the limb of the Moon or Sun and fade quickly with angular distance from it. At most they appear only 1° or 2° in diameter as they are caused by light passing through lower clouds and mist formed of water droplets. If either appears to thicken, the effect is due to thickening clouds than can be in the forefront of rainy or snowy weather; thus, a ring of either kind is sometimes a harbinger of foul weather. Coronas can sometimes appear around Venus or Jupiter because of their brightness.

Sometimes if the Sun is low in the sky and ice crystals are present, whether or not cirrus clouds are also on the scene, only part of a halo can be viewed. If so, the directions where the halo is brightest are the directions directly to the left or right of the Sun and at the same altitude above the horizon. These are the directions of thickest cirrus or ice crystals, and are therefore most noticeable. Pieces of a halo in those spots are known as sun dogs or parhelia.

Halos, coronas, rainbows, fogbows, sun pillars, and mirages come in assorted sizes and flavors. Entire books are devoted to them, usually with color photography to show their variations. But Grandpa's mirage must, I'm afraid, be relegated to his fancy.

✳

15

On a Starry Night

The mind's deepest desire, even in its most elaborate operations, par-
allels man's unconscious feeling in the face of his universe: it is an
insistence upon familiarity, an appetite for clarity. Understanding the
world for a man is reducing it to the human, stamping it with his seal.

Albert Camus, *The Myth of Sisyphus*

On a starry night, a very clear, dark, moonless night, do the stars look as
if they are hanging down closer to us, as some writers would have it, or
infinitely far off, but much greater in number?

Our two eyes provide us with depth perception, and we learn from
experience at an early age that some objects are closer than others
because they appear to be more displaced at wider angles when viewed
from one eye after the other. The distances of familiar objects, such as
trees, houses, or automobiles, we can judge from experience because we
know how large they really are. But from a long way off, hundreds of feet
or more, we can no longer distinguish relative distances among unfamiliar
objects. Objects in the heavens—clouds, for example—are almost impossible to gauge in linear size because we have seen few if any close up.

The Moon and a distant star look to be at the same distance, despite
the ratio of many billions to one in their true distances. It is simple to reason that all celestial objects lie at the same distance because no one can
assess their actual size from their appearance. Perhaps that is why many
inexperienced observers describe the distance between two stars in

inches or feet. Most people are unused to thinking or estimating in angular units but we must if we have no preset size of an object. Airplanes appear to subtend many different sizes of angles, depending on their sizes and distances, but they are perceived as of a certain linear size. The Moon and the Sun always appear about the same angular size, but could anyone guess simply from seeing them that the one is four hundred times the diameter (and therefore the distance) of the other?

The Sun and the Moon appear to be immense when rising or setting compared to other moments when they are riding high in the sky. The same goes for constellations; from our midlatitudes, both the Big Dipper and the familiar W of Cassiopeia are circumpolar; that is, they never set. Each in turn circles up nearly to the zenith and down nearly to the northern horizon. In the latter mode, each looks to stretch along the horizon forever, but aloft they are not impressively big. We view the heavenly vault by day or by night not as a spherically shaped hemisphere but as an umbrella, shaped more oblately with the sky at low angles seeming much farther away than the sky high overhead. No one is quite sure why this is, but the perception is apparently universal among humans, and maybe other animals as well.

As mentioned earlier, the apparent brightness of a star or planet is impossible to retain in the memory. We see any bright star at a great range of magnitudes, from faint under hazy or light-polluted skies, to brilliant when viewed under near-ideal conditions. Yet we think of it as at a constant brightness level. Colors of stars are also deceptive under varying conditions. In a clear summer sky, the two brightest stars are seen in living color; Vega is blue and Arcturus appears orange. Among the next three, Altair and Deneb can be seen as blue like Vega, and Antares is of a reddish tint. The stars fainter than these zero- and first-magnitude objects are rarely, if at all, seen to be other than white. Here perception emulates reality. The cool stars Arcturus and Antares are indeed red or orange in color, while the other three are hotter and therefore appear blue. The rods among the light-detection cells in the retina of the eye are receptive to faint light intensity related to scotopic vision, more than are the cones of photopic vision, but the cones can detect color. For this reason a star's color is undetectable long before it is too faint to be seen at all.

From all of this it is not hard to understand the view that prevailed in the ancient and medieval worlds that held the sky to be a single celestial sphere with all of the stars plastered onto it and thus all at the same distance. There is no depth to the sky, and only because they are seen to move are the Sun, Moon, and planets figured to be closer; the faster the apparent motion in the sky, the nearer the object. The Moon moves most quickly and Saturn the least among the naked-eye planets and the Sun and Moon. Occasionally the Moon is seen to occult one or another planet, confirming it as our nearest neighbor, with Saturn out there just this side of the stars.

The medieval model, in Europe and elsewhere, considered the heavens to be filled with a pure crystalline substance, known as the quintessence, unlike the four elements that were thought to form the substance of our globe: earth, air, fire, and water. It was assumed to be a plenary cosmos with no empty spaces and within which the Sun, Moon, planets, and stars rode around the world in circular paths, as the circle was the only perfect form. The world, though relatively small, was the center of everything and the vast but finite universe was appropriately awesome without being totally incomprehensible as infinity would have been. This universe was old, but only old enough to fit the generations mentioned in the Bible, some several thousand years, again a long but wholly imaginable period of time. Orderliness and constancy were the keynotes of that static world, invented and policed by the deity. Forward progress, so much an innate part of our times, is a concept only a few centuries old. Before then people held little expectation or hope that tomorrow would be better, cleaner, healthier, or richer than today or yesterday. Wars and plagues came and went with no end seen or expected.

Then came the most momentous sea change in history. With less certainty and proof than global warming has today, creation became heliocentric, the Earth was shoved into the planetary lineup, and the stars had to be divorced from the solar system altogether, as they had of necessity to be other suns, not planets. Light pollution aside, we see the very same night sky that they did; the changes were intuitive, not corporeal, and they came about mostly before the invention of the telescope, although that instrument helped to confirm the new reality. Reality in celestial matters, at least, had broken loose from perception and common sense once and for all.

Around that time, Mercator and others realized the impossibility of mapping our round world onto a plane. At and near the tangent point of a plane, the point at which it touches the globe, the projection is reasonably good, but well away from that point it is distorted beyond the useful. So is it for the heavens; we cannot project the familiar constellations onto a flat surface except over a small region. The concept of the sky as the inside of a sphere, though totally incorrect as a model of the three-dimensional universe we know, is nonetheless useful and practical as a concept for surveying and navigation, and it still registers to our sight as a kind of reality. As star maps became more and more common, it was and still is sensible to portray the brighter point sources of light, whether planets or stars, as larger dots on a map, and indeed our eyes and other optics see them as larger in diameter. With magnification the planets are quickly resolved into disks, but no star other than the Sun is ever seen as other than a point, even if observed through the largest telescopes. In fact, this was a key argument in the acceptance of the Copernican system over the geocentric ones. As telescopes got more powerful, planets appeared larger but stars remained points, enhancing the generic difference between them. This had the effect of removing stars to such distances that the failure to observe their parallax, their reflection of the Earth's orbital motion about the Sun, was correctly laid to the distance being too far for the parallactic motion to be detected, rather than to evidence for a stationary Earth.

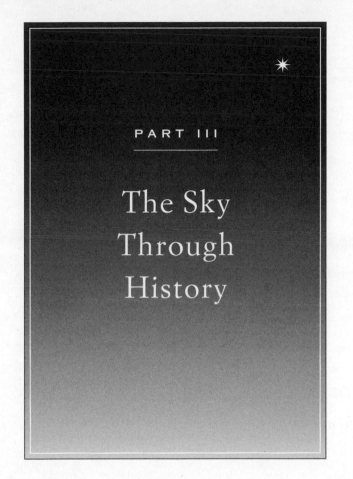

PART III

The Sky
Through
History

16

✳

A Green Mars in the Sky?

There are more things in Heaven and Earth, Horatio, than are dreamt of in your philosophy.

Shakespeare, *Hamlet*

Two conflicting statements appear in the media from time to time. In one, the claim is made that half of all of the human beings who have ever lived are alive today. The other places this figure at about 10 percent. Whatever the statements' origin, the gross difference between these two numbers is of such a magnitude that two widely diverging hypotheses seem necessary to explain them. Surely a model could be fashioned from the available data that would disprove one or the other (or possibly both) of these estimates.

Three sources provide the information necessary to construct a simple model of the human population throughout history: *Atlas of World Population History* by Colin McEvedy and Richard Jones (Penguin Books, 1978), *Four Thousand Years of Urban Growth* by Tertius Chandler (St. David's University Press, 1987), and *The World Almanac and Book of Facts 2002* (World Almanac Education). At the outset several assumptions must be made, the first being the time the count should begin. (Anthropologists date the appearance of the first Homo sapiens to be sometime near 300,000 B.C.) Then, the human population of the world must be reasonably well known throughout the period since that time. This number seems fixed within acceptably narrow limits. Fortunately, moderate-sized uncertainties in either assumption do not alter the result significantly.

Finally, the shakiest necessary piece of information is the average human life expectancy at any period. This must be known or assumed in order to estimate the rate of replacement of one person by another. In historic times, in fact not so long ago, the average life expectancy was very short, only some 20 to 30 years in much of the world. Now it is well over 50 years. Thus at one time in the past, five different people filled one slot in the course of a century, whereas at most only two do so today.

The world population appears to have grown at fairly constant rates over long time intervals until modern times. We are free to assume an average population over an interval, which is simply the population at the middle of that period. We are well aware that the world population has exploded in recent times, tripling in the twentieth century alone. The billions alive today would seem to overwhelm the early years when far fewer of us walked on Earth. That half of all Homo sapiens are alive at the moment would seem plausible. This appears especially the case as we witnessed the great urbanization of the world throughout the twentieth century.

The populations of urban areas past and present have been estimated as carefully as possible. They show that Rome was probably the first city and metropolitan region to approach or exceed a population of 1 million, which most likely occurred in the second century of the Christian Era at the height of its empire. The next cities to exceed or come close to this mark were Baghdad and Constantinople (Istanbul) in the ninth and seventeenth centuries, respectively. Beijing definitely did so by 1800, and by 1900 sixteen urban areas are estimated to have surpassed the million mark, and four (London, New York, Paris, and Berlin) were in excess of 2 million. Now, a century later, more than one hundred cities exceed 2 million in population, and hundreds more exceed 1 million. But, impressions to the contrary, the earlier periods lasted over many millennia with a much higher birth rate, and they add considerably to the total population.

Table 16.1 shows how the total population has been derived. It is not intended to be a definitive study of demographics, but rather illustrative of a model adequate to respond to the two widely different hypotheses mentioned above. It is divided into intervals of time such that within each interval, the population can be assumed to rise at a nearly constant rate. In the second column is shown the world population at each date

TABLE 16.1

The Estimated Number of People Who Have Ever Lived

Epoch	Population Average (Millions)	Weighting Factor	Life Expectancy	Number of People (Millions)
300,000 B.C.	0.0			
		1.0	20	10,000
100,000	1.7			
		2.5	20	11,250
10,000	4.0			
		4.5	25	900
5000	5.0			
		6	25	240
4000	7.0			
		10	25	400
3000	14			
		20	30	670
2000	27			
		35	30	1,160
1000	50			
		100	30	3,330
A.D. 1	170			
		200	30	6,670
1000	265			
		360	30	6,000
1500	425			
1550	485	485	35	1,360
1600	545			
1650	580	580	35	1,530
1700	610			
1750	720	720	35	2,000
1800	900			
1850	1,200	1,200	40	3,000
1900	1,600			
1925	2,000	2,000	45	2,200
1950	2,500			
1975	4,000	4,000	60	3,330
2000	6,000			
Total				54,040

listed in the first column. Next follows the population for the middle of the epoch extending between the two adjacent dates shown in the first column. The remaining data are similarly staggered and apply to the whole interval. Thus the subsequent column gives a weighting factor, which is the number of centuries within each time period multiplied by the average population within that period. During the last few thousand years the curve rises and the population at the midpoint of an era can be read directly off a graph showing the world population over time.

Different sources agree fairly closely with these data for the world population throughout human history. We have now a weighting factor that is proportional to the number of people who have lived on the assumption that each lived for 100 years. This number must be divided by the fraction of a century for the average life expectancy. The largest source of error in the table probably lies in the estimates of the life expectancy, shown in the next column. Only for the last century or two is that figure well known, and the contribution from infant mortality may or may not have been properly taken into account at all times in the past. The final column shows the numbers of lives, in millions, that occurred within each period of time. Since the average lifetime is taken as 20 years, rising through historic times to 30, 40, and today over 50, the population contributions of the first two intervals can be seen to dominate the final result.

According to this very simple procedure some 54 billion of us have ever lived (through A.D. 2000) and about 6 billion are now living, or just about 11 percent of the total. If, instead, we were to begin the count at 10,000 B.C., just after the end of the last ice age and around or shortly prior to the discovery of agriculture, the total is 33 billion, of which 18 percent are now alive. This date is near the end of the Pleistocene epoch and the start of the recent or Holocene epoch. It marks a convenient time after which all members of our genus, Homo, were unquestionably members of the single species, Homo sapiens. A further justification for this later starting date is that it was a time of transition shortly after migrating tribes had settled North and South America, and not long before the first signs of permanent settlements and other auguries of civilization (such as a written language and metallurgy) may have started to appear.

In either event, the great recent population explosion, being of very brief duration, contributes only modestly to the aggregate, overwhelming though its numbers appear, and certainly does not approach the requirements of the 50 percent hypothesis. This and the present longer life expectancy lead to the rather small percentage of the total accounted for by the living. Clearly, the hypothesis holding that half of our species are now living is grossly incorrect. Underestimation of infant mortality in early times would only serve to weaken this hypothesis further.

Most projections of the world population to the year 2050 are optimistic and call for a leveling of the world population to a number near 9 to 10 billion at that time. In late 1998, the United Nations revised its population projections downward slightly due to a continuing decline in the global average fertility rate and an increase in the toll of HIV/AIDS in the underdeveloped countries. The United Nations prediction for 2050 is now 8.9 billion for the world population. If these projections are correct, they indicate that 15 percent of all Homo sapiens will then be alive, or 23 percent of those born since 10,000 B.C. We must hope that we do not exceed this amount by much, for if we do not soon curb the expansion of our numbers by intention, it will be limited for us by one or another of the four horsemen of the apocalypse, or perhaps by a new fifth member of that club in the form of an ecological disaster.

If the population at the next midcentury remains near 9 billion and the life expectancy settles near 65 years, each century that passes will add some 13.5 billion more of us. By A.D. 2400 we would about double the grand total by adding another 54 billion. The percentage of the living would have fallen to 8 percent and would continue to drop unless the population increased once more, as it surely will if we colonize the solar system and the Galaxy.

※

COLONIZATION OF ANOTHER world lies so far outside our present capabilities that it may be difficult for most of us to comprehend that the means by which this would be accomplished are, in a general way, already understood. Lately considerable effort has been expended over the study of the possibility of the terraforming of Mars, the conversion of that planet to one much more like the Earth and less hostile to human life, and the methods by which this far-flung goal might be attained. This is not altogether a pipe dream when we think in terms of centuries. This always assumes, of course, that we do not blow ourselves and our civilization apart in the meantime.

The surface of Mars measures about 55.7 million square miles, almost exactly the land area of our world (57.9 million or, if Antarctica is excluded, 52.5 million square miles). A successful terraformation could possibly sustain as many folks as does this globe, unless we also were to colonize Earth's oceans in some manner, in which case the total surface area of our larger planet rises to 196.9 million square miles.

The Moon differs more from the Earth than does Mars, and might appear to be more difficult to be made to resemble it, but the Moon has the offsetting advantages of being smaller than Mars and much closer, less than one hundredth the distance to the red planet at its closest. The Moon contains only 14.7 million square miles, making its surface area a little larger than Africa, our second largest continent, with 11.7 million square miles. It is hard to imagine that any additional planet allows any likelihood of terraformation, even when compared to the Moon and Mars. Mercury and Venus are closer to the Sun and just too hot, with daytime surface temperatures in excess of 700°F, and Venus is spoiled further by a carbon dioxide atmosphere with a fearful greenhouse effect and a surface pressure ninety times that of our world. If all of our oceans boiled off into steam, we would only then have an atmosphere as dense as that of Venus.

The great gas giants (Jupiter, Saturn, Uranus, and Neptune) are far too large and hostile, and even their larger satellites are probably much too cold to sustain an Earthlike temperature, although they do possess abundant quantities of water in the form of ice. They and all of the rest of the solar system make poor real estate by any measure. Despite our vivid science fiction on the subject, the stars are of such distance that visitation, let alone colonization, is unrealistic at best, even if suitable planets were to be discovered around some of them.

The combined land area of these three worlds—the Earth, the Moon, and Mars—amounts to 128 million square miles, and our oceans raise the lot to 267 million. Hence, we could in theory eventually settle over three to five times the surface area now available to us. It may or may not be absurd to anticipate that the number of living Homo sapiens may quintuple to some 45 billion people by the year 2500 or 3000.

The terraforming of Mars or the Moon must assume that a large quantity of water exists or can be economically generated or transported there. With water, the atmospheres can be, at least in theory, intensified to pressures closer to ours, with leavening temperatures and some oxygen present. Such a conversion of Mars could, in our wildest dreams, transform that planet from a reddish dusty desert world into a green and blue one like the Earth. The Moon, too, might take on a similar cast, and the brightness of both objects would rise considerably, since their thicker, cloudier atmospheres would reflect more sunlight, rather as Venus and the Earth do now. As mentioned earlier, the percentage of incoming sunlight reflected straight back into space, called the albedo, is low for worlds with thin or nonexistent atmospheres, being some 10 percent for the airless Moon and Mercury, and only 15 percent for Mars. The partly cloudy Earth reflects about 37 percent of its light, and the albedo of the totally cloud-enshrouded Venus is twice that or about 65 percent. Rife with verdure and with a higher Earthlike albedo, Mars could at rare times be bright enough to be seen rising into the daytime sky shortly before sunset, as Venus can be seen in the daytime now. An oxygen-rich full Moon would be so luminous that only the brightest stars could be seen on a clear night.

About a century ago, when the true nature of the Martian surface was first sensed, a proposal was made to carve a diagram of the Pythagorean theorem (see Figure 16.1) into the northern taiga, the great coniferous evergreen forests of the subarctic regions. The theorem requires the sum of the squares of the two sides of a right triangle to be equal to the square of its hypotenuse. This figure would extend for hundreds of miles and would have the purpose of providing any intelligent race on Mars or any other planet with proof that intelligent life also flourished on Earth, when they glimpsed it through their telescopes.

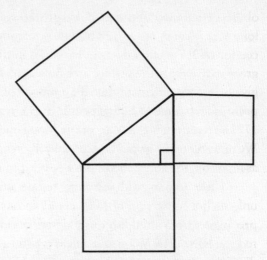

16.1 The Pythagorean theorem. The sum of areas of the two smaller squares are always equal to that of the largest square along the hypotenuse.

Needless to say, the project was never undertaken, doomed by (aside from its frivolity and our natural procrastination) the development of space flight. First, the idea of actual visitation presented itself as a better way of contacting another intelligence. Second, the first images of Mars from a nearby space probe disproved the theory of Percival Lowell and others that a network of straight lines crossing the deserts there were actually canals on Mars and stood as proof of a high-level civilization, whether now extinct or not, that sought to irrigate its deserts with water from the melting polar caps during the Martian spring, when they showed that no such lines exist.

We have learned much about the climate of Mars and the changes it is expected to undergo over time, since the *Viking* landers settled down onto its surface in 1976. As mentioned earlier, the planet has an atmosphere with a surface density just 0.7 percent of ours. This thin air is about 95 percent carbon dioxide. If this were all we could say about it, the chances for any form of life there would be slim indeed. But a number of surface features indicate water erosion; at some point in its distant past, Mars may have had open water on its surface. To understand how this might be possible, we must return to the very slow changes in the obliquity and orbital characteristics of a planet. Milankovic's theory tied long-term variations in our climate to these slow changes. The first is precession, the 26,000-year wobble of the Earth's axis, brought about by gravitational attraction of the Sun and Moon on the equatorial bulge that gives the planet a gyroscopic motion of its axis. Sufficient time has passed for the determination of the precession of Mars; it is about 177,000 years, almost seven times the length of precession here on Earth. We expect that because Mars has no large moon and the Sun is half again as far away.

More to the point, the tilt of its axis goes through wild gyrations, unlike ours, with its modest variation of about $1\frac{1}{2}°$ more and less than its present $23\frac{1}{2}°$ inclination, with a period of around 41,000 years. On the red planet this motion is chaotic, meaning that it has no regular predictable or cyclic variation, but vacillates widely from about 15° to 35° in tilt from its orbital plane, its ecliptic so to speak, which is inclined only some 2° from ours. Its present value is 25.2°, very close to the Earth's, but it can vary by as much as 13° above and below that value over 100,000 years or more. On the Earth, a low tilt tends to bring colder temperatures, possibly leading to a new ice age, because the Sun's lack of altitude in the sky at the higher latitudes equates to cooler summers and less snowmelt. A high tilt allows the Sun, high in the sky, to melt rapidly last year's snowfall. Translated to Mars, a tilt of 15° leaves the air so cold that almost all of the Martian atmosphere freezes out on the ground, whereas at 35° or so, the relentless summer Sun melts so much that it makes for an atmosphere much denser than its present one, with water being one of the probable constituents. So water may have flowed in rivers on the surface several times in the distant past. If this is the case, the chance for some type of life there now or more likely much earlier rises dramatically. This water is now lodged in the two polar caps and in the Martian soil beneath its surface. Terraformation would be much less difficult if these theories and observations are correct. Further space ventures to Mars are planned with just this goal in mind.

Still any speculation on the subject of the terraformation of the Moon or Mars borders on the whimsical at best. We may have grown wiser in this century, or at least since 1912 when much of our collective hubris went down with the unsinkable *Titanic* in that most infamous of disasters. We are less likely now to accept that nature is here for us to use and misuse to our own satisfaction. Whether or not we ever do settle and colonize other parts of the solar system cannot now be predicted. But if our race should succeed in populating these other worlds at some remote time, might it not be too optimistic to assume, in that brave new solar system, the eventual presence of another musical genius of a level comparable to Mozart?

1 7

✳

Astrometry and Creationism

The great tragedy of science—the slaying of a beautiful hypothesis
by an ugly fact.

Thomas Henry Huxley

Not long ago it was said that the concept of the universe being fashioned by a creator only a few thousand years ago violates every major
subsection of astronomy except astrometry. But astrometry, the science
of measuring the stars, particularly their positions and apparent motions
over time, called proper motions, is also at odds with this obsolete cosmology. The best known of these instant universes is that of the Irish
bishop James Ussher; it is one of several variants conceived around the
seventeenth century, in which the time of creation was fixed well within
the last 10,000 years. Bishop Ussher placed the creation on the morning
of October 23 in the year 4004 B.C. Ussher's timing of the event is the
most widely quoted, probably because it is so precise, even if inaccurate
(a clock, fast by an hour, may be precise to the minute, but it is still inaccurate by a whole hour).

Most creationists, of his time and today, subscribe to a time close to
Ussher's for the formation of the world, the solar system, and the whole
of the universe, based largely on the account given in Genesis. But soon
afterward, this façade developed major fault lines. The trouble was fos-
tered by the confirmation of the heliocentric arrangement, first that of
Copernicus and later of Kepler and Newton. This new theory demanded

distances to the stars that are not just beyond Saturn, then the farthest known planet, but many, many times as far, since their parallactic motion, now known to exist, could not then be detected. Many sought this back-and-forth swing that reflected the orbital motion of the Earth about the Sun, but no one could detect it until 1838. In that year, three astronomers, working independently of each other, all measured the first accurate distance to a star. Clearly the sophistication of telescopes had enabled them to make the measure that had not been possible during the previous two millennia.

Motion governed pretelescopic astronomy. Only the Sun and the Moon could be seen as more than points of light; no astrophysics could apply to visual inspection of a point of light and astronomy was limited to astrometry, especially for the stars. A small universe had been adequate for Ptolemy and the other Hellenistic astronomers, one barely the diameter of our orbit around the Sun, about 200 million miles across, twice 93 million miles to be precise. These dimensions took account of the distance to the Sun, brilliantly deduced but inadequately measured by Aristarchus in the third century B.C. He held that the Sun was about twenty times as far and hence as large as the Moon, or some 5 million miles away. The true value is twenty times 20, or four hundred times the size and distance of the Moon.

In Newton's and Halley's time, a number of techniques placed the Sun roughly as far as we now know it to be. About then we also became aware that the speed of light, although immensely great, was not infinite. In 1729, James Bradley, who succeeded Flamsteed and Halley as Astronomer Royal, dealt the first of two final fatal blows to geocentricism, long since abandoned by his forebears for reasons just short of absolute proof. By detecting the aberration of starlight, he proved that the speed of light was not infinite and our orbital speed was not zero. Only the first confirmed stellar parallax a century later was as irrefutable a confirmation of the moving Earth as this. This parallax is the angular size of the Earth's orbit as would be seen from the star. From any star, the orbit would appear as a tiny ellipse, and the farther the star, the smaller the ellipse. The relative distances to the planets were known, but only around this time did the absolute distance of the Sun, and therefore, by inference, the absolute distance of all the planets, approach modern values.

The aberration of starlight is properly likened to driving at night through a windless snowstorm. As the snow falls vertically, the car moves horizontally and the snowflakes are seen in the headlight beams to radiate not from the zenith, but from a point ahead of the zenith in the direction of motion. If the car increases its speed, the snow seems to come from a point farther ahead. If the car turns, the snow appears to deviate in such a way that it remains coming from directly in front of the moving car. Were the car to move in a circle, the snow would be seen to radiate from directions that also form a circle. If the car were to stop, the flakes would appear to fall from the zenith, and should the snow cease to fall and instead remain suspended in the air, they would appear to come from the point ahead toward which the car is moving. Coming from an intermediate position proves that both the car and the snow are moving relative to the ground.

So it is with light—if the Earth were at rest, the light source (star) would appear stationary in the sky. If the speed of light were infinite, the same effect would be seen. But stars all move in tiny ellipses around a fixed position at a radial distance from it of about 20 seconds of arc, placing the Earth's orbital velocity at about one ten-thousandth that of light, and both move at speeds that cannot be either zero or infinite. Bradley had found the first absolute unequivocal proof of our motion around the Sun. The first parallaxes were detected about a century later, as noted above, because they are all less than 1 second of arc, a much smaller angle.

Scientists had known about how far the nearest stars must be, since they were known to be suns about as bright as our own, but not until we could finally detect the apparent sizes of the tiny motions stars undergo due to our own orbital motion could we gauge the star's distance with certainty. The nearest of them turns out to be over seven thousand times as distant as Pluto, our farthest planet, itself forty times as far from the Sun as we are. The light-year was developed to circumvent the huge numbers otherwise necessary and this fixed the concept of time as a measure of distance. Soon it became known that some stars were hundreds and even thousands of light-years away; light required centuries and millennia to make the trip to the solar system. The geological evidence favoring the Earth to be far older than Bishop Ussher believed was soon supported by a counterpart in stellar astronomy.

The nineteenth century was marked by equivalences determined between angular and linear motions of ever larger, brighter, and more distant objects. Distances were derived that taxed and even exceeded the 6,000 years allowed by a strict interpretation of the Bible. By 1859, when Charles Darwin published his seminal theory of evolution of the species, light was known to require tens of thousands of years to cross the Milky Way, our galaxy. In the 1920s, Harlow Shapley, Heber Curtis, and others derived distances and hence light-times of hundreds of thousands and even millions of years to parts of our cosmos.

Then came Edwin Hubble and his discovery of a huge and expanding universe billions of light-years across and thus billions of years in age, and in 1952 Walter Baade and his recalibration of it left it twice its former accepted size, and hence unequivically much older than our 4.6-billion-year-old Earth and solar system. A steady progression of well-calibrated methods of the comparison of distant and nearby objects with the same spectral and physical characteristics, and of setting linear against angular sizes and distances of stars and later nebulae and galaxies, has been made throughout the last 50 years or more. They have ended any possibility that the systems of stars forming our local galaxy and the systems of galaxies defining the observable universe are less than billions of years in age, far older than any schema derived from a literal interpretation of the Bible.

All methods for the measure of cosmic distances outside the solar system ultimately depend on matching a linear with an angular dimension. A firefly, an airliner, and a satellite in near-Earth orbit may all appear to move at the same angular velocity across the heavens. But if their linear speeds are known, each can be set at its rightful distance. This is a rule that can be made for a group of objects near each other in the sky, even when the distance to no single member of the group can be determined. With the development of the spectroscope and later the spectrograph (in which the spectral appearance of a star can be seen and photographed, respectively), astronomers became capable of measuring an object's motion in the radial direction, the direction along the line of sight between the object and the observer. But having measured this, we can say nothing about a single star's lateral or sideways motion in the plane of the sky. One star may move more or less directly toward or away from us, in which its lateral motion is relatively small. Another star may move mostly laterally. But when we consider a group of stars, the situation changes. Gravitational interplay between these mutual neighbors will quickly smooth out any favored direction left over from their creation. Therefore the total distribution of the motions will be about the same in the radial direction as in the lateral direction. We know then at which distance the apparent proper motions across the sky equal in size the spread of radial motions. Years ago, the motions of other galaxies were incorrectly known. Scientists had assumed too small a distance and therefore their radial velocities appeared to be much larger than their sideways motions. If that were the true case, the direction of a galaxy's greatest motion would depend on the direction of it from our own galaxy. All large motions would point directly toward or away from—us. We know we are not so favored, and these "fingers of God" all pointing at us were not real. Thus does science advance.

Some creationists still contend that the Earth and the universe were created in 6 days some 6,000 years ago. They dispute the findings of astrometry, astrophysics, and geology as imaginary, placed in our collective memory by God. Perhaps God gave us so many ways to determine the greater size and age of the cosmos to test our faith in Genesis, but to accept that biblical account as true now requires the rejection of science and its ability to measure things altogether. This argument is just as specious as another that might hold that God created us all last Tuesday afternoon complete with memories of earlier times in our lives, our parents and others, and the history of our species and our world. Certainly, the creationists will develop no credibility as long as they push only the account in Genesis over all the many other creation theories put forth by non-Western theologies from every continent but Antarctica, for with Genesis their true motives of pressurized proselytization for its own sake become clear.

Either way, this makes a meaningless game of the origin of the universe that anyone can play, and the more gaudy and absurd one makes the world, the greater attention one is likely to receive. The quest for knowledge becomes as silly and irrational as it would with the literal acceptance of Santa Claus making toys at the North Pole, perhaps with the assistance of the tooth fairy and the Easter bunny. Galileo and Darwin present two of many cases where science and religion had been mutually confused. Why can we not agree that their domains can coexist separately once and for all?

Lately we have seen one more example of this foolishness. In 2001 at the time of this writing, the entire human genome was charted with the many thousands of individual genes in place. The development promises great medical advances in the years to come. Relief from cancer, Alzheimer's disease, AIDS, cystic fibrosis, and many other ills of our species may at last become a reality. Yet 40 percent of Americans polled at this time claimed that the genome-mapping program was immoral. If God had wanted us to fly, he would have provided us with wings, these people once claimed, never stopping to ponder the fact that since he provided us with brains, he must have wanted us to think.

18

✳

Those Universal
and Unalienable Laws

We hold these Truths to be self-evident, that all Men are created
equal, that they are endowed by their Creator with certain unalienable Rights, that among these are Life, liberty, and the Pursuit of
Happiness.
Thomas Jefferson, The Declaration of Independence

It may appear perplexing and contradictory that Isaac Newton, of all
people, should feel the need to dabble in the inexact science of alchemy.
This is particularly so if he felt self-motivated or pressured to take leave
from the natural sciences with their universal laws that he, more than
anyone else, formulated and brought to the consciousness of the educated world. He clearly became the central persona of the greatest of all
revolutions, of all sea changes in human thought.

Many examples of diversion from such straight and narrow allegiance
occur. As noted in an earlier chapter, miracles as departures from those
laws are among the most pervasive examples. But, as Barbara W. Tuchman so eloquently makes clear, we have a changed world from the one
that held sway in Europe before the scientific revolution that culminated
with Newton's laws of motion and gravitation. The impact of their universal nature was capped by Edmond Halley, Newton's longtime colleague. Halley used Newton's laws to prove that comets follow orbits

around the Sun and are therefore not of this world. Ethereal as they may appear, they are in no way to be classified with rainbows, mirages, and other aspects of our own atmosphere, or with angels, ghosts, and other creatures of myth.

Halley noticed that the bright comets that appeared in 1531, 1607, and 1682 followed the same orbit and thus were the same comet, the one now named in his honor. He then made the audacious prediction that it would again appear as a spectacle in 1758, some 76 years in the future, and indeed on Christmas Day of that year it was spotted just where he said it would be. The triumph of this feat and the meaning of it impacted the entire intellectual world community, not excluding the Founding Fathers of this republic in their salad days. Bernard Cohen (in *Science and the Founding Fathers*) has traced the direct lineage from the universality of these laws to the unalienability of the truths on which they founded their new nation in 1776.

＊

MANY AMONG US make exceptions to these universal laws of motion to defend a particular viewpoint. This is the kind of incident I mean: Can anyone really imagine that Newton's third law of motion was somehow inactivated in Dallas on November 22, 1963? The law states that for every action there is an equal and oppositely directed reaction. The evidence pertaining to the law is in the film made by Abraham Zapruder of the assassination of President Kennedy. At eighteen frames per second and viewed frame by frame, the film shows clearly (in frames 313 to 320, less than half a second) that the third and fatal bullet moved the president's head backward. The Texas Schoolbook Depository was situated behind the car in which the president was riding. Thus Lee Harvey Oswald, allegedly in that building, could not have fired the lethal shot; it came from the front (released autopsies confirm that the entrance wound was on the forehead). Neuropathologists confirm that no physiological or muscular reaction could, in the time between one frame and the next, account for that backward recoil. The action had to and did come from the front. It is essential to keep in mind here that physics goes no farther; it has nothing to say about a gunman on the grassy knoll or elsewhere, or about a conspiracy between Oswald and anyone else. It says only that the fatal bullet came from the front, no more. Had Oswald lived and gone to trial, the whole mythos surrounding the assassination and subsequent events might have taken a very different course of action. Newton's laws are universal or they are not; no solipsistic or convenient lacuna can be substituted for this reality. (Here we take a still-controversial example advisedly; the point is not so clearly made using an example such as a flat Earth, which has long since been rejected by everyone.)

It is altogether too rarely expressed in texts on astronomy or astrophysics that the very universality of Newton's laws connects the macrocosmos of stars and planets and the microcosmos of individual people and animals. Their motions seem differently governed, but they are firmly joined and form a single continuum in ways that are not readily apparent. Only as one probes into the subatomic domain do these rules appear to change. There the uncertainties of quantum physics take over, and motions behave differently.

But let us explore our larger arenas of action. As stated earlier, the Moon's gravitation is much smaller than the Earth's. The velocities needed to escape the Earth or the Moon, although expressed a little differently than the causes of weight, also depend on the masses and distances involved. A velocity of 7 miles per second is necessary and sufficient for an object to leave our world for distant space, no matter how the velocity is achieved. From the Moon, only a 1½-mile-per-second push is required to escape. What about even smaller worlds? They, too, follow the same rules. Ceres, the largest and first discovered of the multitude of asteroids circling the Sun between Mars and Jupiter, is only about 600 miles in diameter, less than one third that of the Moon. There, our 180-pound man would weigh only 7 pounds, as opposed to 30 on the Moon, and with a decent arm, he could throw a baseball for about a mile.

Let's go even smaller. Mars has two tiny moons named Phobos and Deimos, about 10 and 5 miles across, respectively, and shaped rather like potatoes. Objects much smaller than Ceres are not globular or even

146 · ARTHUR UPGREN

close, because their own gravitational fields cannot overcome the rigidity of their material. As they formed and cooled, so are they yet shaped, and no rondure or "roundifying" process could take place.

Nonetheless on tiny Phobos, a game of baseball must be played with a new and rather different definition of a home run. There a home-run ball would indeed go out of play, for it would leave Phobos and proceed in its own orbit around Mars. If we move the ball game to a yet smaller world, one perhaps 1 mile in diameter, the approximate size of most newly discovered asteroids, the outfielder jumping to try to snag a long fly ball would follow it right on out into the void. Each team would need a bull pen of fielders along with the pitchers, at least until some space shuttle could round up and bring back the players from their own orbits in space. Things could be worse; on Jupiter, were it to have a solid surface, pitchers could throw curves but no fast balls and the poor 500-pound batters could never hit anything out of the infield.

No difference in kind is found between the game on Earth and one on a small asteroid, despite the bizarre playing conditions on the latter. Taking this one step farther, we can now understand that the astronaut on an extravehicular space walk outside the space shuttle is still gravitationally bound to it and influenced by it. But in point of fact, the slightest push from it exceeds its puny velocity of escape and would propel the unfortunate astronaut off into space; hence the presence of an umbilical cord tethering him or her to the spacecraft.

It is of interest to many to contemplate the counterfactual in history, the what ifs that can be and have been imagined and explored by historians. What if Hernán Cortés had been killed in 1519 and did not overthrow the Aztec empire? What if the lost Confederate order had not ended up in Union hands and Robert E. Lee had carried the day at Gettysburg? What if Alexander the Great or Martin Luther had never been born? It is widely understood that the great-man theory in history has faults and virtues, and the projected nonexistence of this one or that one makes for speculation on how much his absence would have altered our world. In this vein, I consider the case in which the premature and frail infant Isaac Newton had died in his first week of life in December 1642, and not at almost eighty-five years of age in 1727.

The giants, as Newton referred to his scientific predecessors on whose shoulders he felt he stood, above all Johannes Kepler and Galileo Galilei, had left a great but unfinished legacy. Kepler had discovered that planets moved in elliptical orbits about the Sun, implying that some kind of solar force kept them there, and God had not necessarily started a giant clockwork system of his own design.

Everything Newton accomplished would have been done by others, not likely in such a compact and spectacular fashion, but dribbling out in bits and pieces. As it was, he alone developed the theory of gravitation and the laws of motion by which such factors as force, speed, velocity, and acceleration were first understood and correctly defined, and he did it in a very short time. He gave us the physics needed to account for the solar system, and he developed the calculus and other mathematics required for proof. In his book *Philosophiae Naturalis Principia Mathematica* (*Mathematical Principles of Natural Philosophy*), known familiarly as the *Principia,* Newton addressed a whole range of matters that followed directly from his work. He was the first to explain the reasons for the tides, the precessional motion, and the concept that one object did not revolve around the other but, strictly speaking, both did so around a common center of gravity, the last before it could be confirmed by reliable observations.

He had done most of this in his twenties, but didn't publish until his forties in 1687, when Halley encouraged him to do so. Halley was one of a number of contemporary scientists who tried, and eventually might have made these discoveries, and the German mathematician G. W. Leibniz would have and independently did discover the calculus. But the Londoners Halley, Robert Hooke, Christopher Wren, and John Flamsteed, to name a few, were only beginning to see where they had to go when they found that Newton was there already.

Perhaps his greatest accomplishment was the compelling impact that his laws applied everywhere—that they were truly universal. Gone was the Aristotelian concept that this mundane world follows one set of rules with its four elements of earth, air, fire, and water, while the heavens behaves according to other rules and was made of a fifth pure crystalline substance not found on Earth, known as the quintessence. The very same laws applied and the same stuff was found everywhere. Our Earth is not special in any way, and it has no special favored location in space.

Although Thomas Jefferson and his compatriots would have been aware of these principles, perhaps they and many others would not have been so taken by Halley's successful prediction of the return of his bright comet in 1758. Halley might not have had access to the mathematics required to connect the appearance in that year with the three earlier ones. The impression on the Founding Fathers might have been diluted to the extent that their insistence on unalienable laws might not have been so resolute.

Astronomy followed Newtonian mechanics (as well as Newton's seminal work on optics) for more than 2 centuries with one success after another. The discovery of Neptune, the distances to the stars and their motions, and the eventual confirmation that all the stars in our galaxy revolve about its distant center were all children of his mechanistic system. Only Albert Einstein modified and altered the cosmos from Newton's universal model and gave gravity a different basis.

✳

19

The Egg and the Equinox

The only ethical principle which has made science possible is that the truth shall be told all the time. If we do not penalize false statements made in error, we open the way for false statements made by intention.

Lord Peter Wimsey, in Dorothy L. Sayers's *Gaudy Night*

At either equinox, the Sun is on the celestial equator, since these two nodes are the points where that great circle crosses the ecliptic. At either solstice, the Sun stands 23½° from the equator, north or south, as far as it can ever get.

The force of gravitation of one body upon another is always directed straight at the center of the perturbing body. The Earth attracts us directly toward its center; Isaac Newton proved this in the work he published in 1687. For all other objects in the universe the same rules apply, but only the Sun and the Moon among the astronomical objects are close enough or big enough to raise measurable perturbations in the directions of their centers.

Under what conditions does either the Sun or the Moon exert a maximum pull on the Earth or anything on it? They need to be at their closest to us—the Sun at perihelion and the Moon at its closest point, the perigee—in order that their tidal effect is maximized. Since gravitation acts proportionally to the square of the distance, we can calculate the range in gravitational strength for each of them. The Sun can vary by up to 6 percent; that is, at perihelion its pull is 6 percent greater than at aphelion. For the Moon this figure is near 22 percent; it is greater than that of the Sun because its orbit around us is more eccentric than is ours about the Sun.

Of greater significance is the fact that the direction of the force is perpendicular to the horizon only when the Sun (or the Moon) is either at the zenith or at the nadir, the point straight down below our feet. At any other time, the force is directed at an angle to the ground.

Balancing a raw egg on one if its two points is a trick for the steady of hand. It can be done but not easily. Somehow many in the media and elsewhere have come to believe that it can be done only at the moment the Sun crosses the celestial equator at the point of one of the equinoxes. Two presumptions must be fulfilled in order for this postulate to be correct; the first is that at that instant the Sun's gravitational influence must be directed overhead in some manner. If it is not, the Sun will be pulling the egg to the side and the egg would tend to fall over. Now think what the Sun is doing on March 21 or September 23, the dates of the equinoxes. Only along the equator can the Sun ever be seen at the zenith or the nadir on either of those days. And, since this globe of ours rotates, the Sun will only pass through either point instantaneously. Only at that instant can the Sun (or the Moon) help in standing that egg up on its point.

The second presumption is that the Sun has sufficient gravitational force to matter at all. Newton's universal law of gravitation states that

$$F = GMm/d^2$$

In words, this equation says that any particle in the universe attracts any other particle with a force that is a product of the masses, M and m, of the two objects times the gravitational constant, G, and divided by the square of the distance between their centers.

Now the Sun is a third of a million times as massive as is our planet, but the center of the Earth is only 4,000 miles beneath us while the Sun is 93 million miles off. We need to calculate the quantity 330,000 divided by $(93,000,000/4000)^2$ in order to measure the relative strengths of the Sun and Earth. This amounts to a puny 0.06 percent, one part in around

1,640, of the Earth's pull, and the Moon's effect is much smaller yet. In any attempt to balance a raw egg, which may not be symmetrical about its own axis, on a surface that may be neither smooth nor level, I must sincerely doubt that the Sun's influence can count for much. And as it goes by the zenith or nadir (if and only if you are on the equator in any event), it will abruptly change. Yet whenever either equinox is approaching, newscasters remind us of the supposed verity of this nonsense. Sometimes the TV weather forecaster, who knows better, will gently refute the claim, but it still comes back as surely as poison ivy to greet the next equinox.

*

ASTRONOMERS LEARN to deal with many people who accept and promote incorrect, paranormal, or even pseudoscientific reasoning and beliefs. Among these are astrology, unidentified flying objects or UFOs, and the Full Moon effect. Phone calls and letters seem to arrive with regularity from those who advocate these convictions, whereas others call for ever more outdoor lighting despite no connection between lighting and crime, or more defense lest Martians take over the Pentagon, or even that the Earth is flat. In dealing with matters we must be courteous but firm in denial. We try to bridge the gap between perception and reality.

Paul and Anne Ehrlich recently described an increasing trend among older scientists to engage in tithing, giving back to society a portion of the education they have received. They know of no better way to serve those who allowed them to practice science than to point out fallacies affecting scientific issues. Carl Sagan, Isaac Asimov, Sir Patrick Moore, David Levy, Stephen Jay Gould, and many other authors have devoted much of their later years to addressing some of the mistaken and sometimes harmful beliefs that take hold of a segment of the populace. Even such disparate notables as Steve Allen and Richard Feynman speak with a single voice on this subject. This is not always a popular move; one of my own experiences suffices to make the point. During the most recent close approach of Halley's Comet in 1986, I was one of many astronomers asked by members of the local media for the best way to observe the celestial event. My recommendation was to get well away from city lights and observe the comet with 7 × 50 binoculars, which are of a size and type well adapted to night viewing. I was unaware that an ambitious astronomer planned a self-serving gala event with my university's large telescopes. The instruments make for poor comet viewing for several optical reasons. They can accommodate only the head in their small fields, and their optics would not in any event allow the tail to shine brightly enough to be seen against the light-polluted night sky of the surrounding city in which it was immersed. The head could be seen as a faint fuzzy star, which may have satisfied those who wanted only to be able to claim that they had seen it. The others were readily disappointed. A representative of the university administration soon afterward admonished me for failing to favor the staged observing over the use of binoculars in a dark sky location. Within a week, both Sagan and Asimov responded to the same question on national television by recommending 7 × 50 binoculars and a dark sky, as I had done concerning the visibility of the comet. The event recalled a favorite quote of one of my mentors that appears as the epigraph to this chapter.

The all-time champion target of the folks who dabble in science must certainly be Albert Einstein and his theories of special and general relativity. Einstein seems to act as a lightning rod for all the kooks who see themselves as the next mathematical genius. For all the attempts to disprove his efforts, the likelihood of one of them being even slightly correct is minuscule indeed. From 1687, when Isaac Newton published his account of the laws of motion and gravitation in his system of mechanics, until 1905 and 1915 when Einstein's two theories first appeared in print, Newtonian mechanics withstood every empirical test. Only in the late nineteenth century when astronomical observations increased greatly in precision and sophistication did a few very small observational discrepancies reveal themselves and these led directly to Einstein's achievements.

In the tabloids, the junk that clutters checkout lanes in supermarkets everywhere, bizarre "facts" are proclaimed in grand fashion. These have included such whoppers as "Woman Gives Birth to Chicken" and "Elvis Is Heard Singing on the Moon." Just as often, the imagined peccadilloes of one or another celebrity have formed the headlines in one or more of them. My one experience with them came first in a phone call one morning from a reporter working for a local newspaper. He wanted to know if I had any plans to blow up the Moon. I mumbled something to the effect that I did not harbor such plans. About two cups of coffee later another reporter from a second local paper called with the same question. I began to realize that something was afoot. Apparently one of these tabloids came out with a headline that stated, "Scientists Plan to Blow Up the Moon!" (see Figure 19.1). After reading the accompanying article I still had no idea of the purpose of these top scientists (in the world of the tabloids all scientists seem to be at the top in some unexplained manner). I explained that no scientist, top or otherwise, would, could, or should explode the Moon into fragments. Not only is this far beyond the capability of the entire human race, and therefore is in no one's plans, but to do such a thing would endanger the Earth, since some large fragments would surely head our way. Even Adolf Hitler's worst megalomaniacal plans to dominate the world never included target practice of this magnitude on the Moon.

19.1   Headline: "Scientists Plan to Blow Up the Moon!" Courtesy of *Weekly World News*

It is another matter altogether to consider nudging an asteroid into a different orbit. As far as is known, the largest asteroid to collide with the Earth was about 6 to 10 miles in diameter. One of this size struck the Gulf of Mexico and the Yucatán peninsula about 65 million years ago. The ensuing worldwide destruction eliminated some three fourths of all living species including all of the dinosaurs. Were we to face another catastrophe of that size, then top scientists and everyone else, including even Congress, would try to move Heaven and Earth, so to speak, and might just do it. But the Moon is far too large for us to budge, much less break apart, and it is not going anywhere it shouldn't.

People in the skywatching business have handled communications from cranks in a variety of ways. One astronomer of my acquaintance responded to any and all of the letters of this genre that he received with a form letter, the essence of which is as follows:

> In the capacity of Professor of Astronomy, I receive a great number of letters similar to yours. If I gave individual answers with full explanations, I would have a full-time job on my hands. Therefore, I am replying with this prepared statement.
>
> The writers of practically all of these letters are quite untrained in the techniques of science. They have no acquaintance with the vast amount of painstaking observation, measurement, calculation, and double checking that a scientist must do in order to establish even simple facts. They appear unaware that scientists are very critical of their own work, and just as critical of the work of other scientists. The letter writers seem unaware that truth is gained, and error eliminated, by this "check and balance" relation of all scientists acting as referees to all other scientists in their field.
>
> These letter writers are certain in their own minds that they have discovered something of great significance. Most claim to have overthrown basic concepts of science and introduced new ones. Very few of them claim to have extended knowledge by building upon what has already been established—they are always revolutionizing—overthrowing something.

Now I am aware that astronomy has been "revolutionized" a number of times in the past. But the revolution has always come from within. It has been accomplished by people who were trained in the older concepts of the science, and because they were so trained, and had critical minds, they were able to correct erroneous items. Before one is competent to overturn major tenets of science, one first must gain a thorough understanding of the tenets to be overthrown so that he can retain the good and discard the bad. Apparently the letter writers consider it quite unnecessary to have such an understanding. They set themselves up as authorities, and they claim to have ways of attaining knowledge, which make instruments and training unnecessary. To me, this method of "discovery" is entirely incomprehensible.

Of all the numerous letters I receive from these amateur revolutionists of science, I have never found two that agreed on anything. Any one of the letters would flatly contradict all the others. Frankly, then, since your letter is contradicted by all the others, and since they all claim that they and only they know the truth, I fail to see how I can regard your ideas as correct. Dozens of other people—all claiming to be just as inspired as you—would say you are wrong. I can only conclude that all of you are wrong. The proof of your error lies in decades, or even centuries, of hard work by many scores or hundreds of careful and critical workers, who had no desire to deceive anyone—least of all themselves—but who wished only to discover the truth. When a major change becomes necessary, it will be made by those who have been thoroughly trained in that system of knowledge, and who are therefore most competent to discover what is wrong with it.

Curiously enough, it was this same astronomer who taught me a valuable lesson; the underrated value of experience in the interpretation of observations. I had seen Mars many times through a telescope, magnified into a sizable disk. Mars is a disappointment compared to Jupiter with its obvious oblateness, its surface features including the "Great Red Spot," and its four very conspicuous moons. Saturn, too, impresses the observer with its striking system of rings surrounding the planet itself.

Only the Moon, with its mountains, craters, and other surface features, rivals the two giant planets when glimpsed through a telescope. Mars and the featureless Venus disappoint by comparison.

Familiar as I was with the Martian disk, I noticed the conspicuous southern polar cap and a few dark markings near the equator of this red planet and pointed them out on nights of public viewing. But this time I was with an expert on areology (the Martian equivalent of geology) who quickly noted the progress of seasonal changes in the size and shape of these dark features. His experience, compared to mine, was like mine to a first-time viewer of Mars through a telescope.

The problem of UFOs, unidentified flying objects, the flying "saucers" of earlier years, arises in this connection. Notice the word *unidentified* as opposed to *unidentifiable*. Many have quite honestly perceived any number and variety of strange unfamiliar sights in the sky, by day or by night. Few are aware that Venus can, at times, be seen in broad daylight, particularly at high altitudes where our atmosphere is thin and clear. Many pilots mistake this "star" for an alien spacecraft.

At night, a whole host of brilliant objects is seen; the overwhelmingly great majority of them are common objects seen in uncommon aspects. The U.S. Air Force, in *Project Bluebook,* a much-maligned study of many UFO sightings, found about 95 percent of them to be readily explainable by anyone familiar with the sky. The remaining 5 percent could not be explained, and together they reveal the impossibility of proving a negative. In order to prove beyond a reasonable doubt that UFOs do not exist, one must contend with every reported observation, linking each to a mundane cause. Strictly speaking, if even one cannot be so explained, the case against UFOs, though very strong, is not confirmed beyond a doubt.

Like others in my profession, I have seen many celestial objects that were unidentified at the time, but none were unidentifiable. The most arresting sight took place in Maine, where a round faint nebulous object appearing a little larger than the Moon moved steadily in a northeasterly direction until lost behind trees. I made note of the time of the incident, and the speed (the apparent speed in angular terms) and direction of the object—mandatory adjuncts to any observation—and waited. A few days later, a Japanese satellite was reported to have dismissed a cloud of gas over the eastern United States in a manner that fit all of the observations and the mystery was over. It was experience, little more, that gave me a leg up, experience in pattern recognition with celestial things: stars, constellations, weather balloons, blimps with advertising lights, icebound wires strung overhead reflecting unseen automobile headlights, tight formations of birds flocking together, and so forth. Experience in not confusing linear with angular dimensions—something may be this many degrees in diameter, but not this many feet across, especially if the object is an unfamiliar one—experience in knowing that all objects disassociated with the Earth (not just the Sun and Moon) share the sidereal motion of the sky from east to west reflecting the Earth's own rotation.

It is not easy to inform some well-meaning person that his or her interpretation of observations is wrong. And when it cannot be done with certainty, it is perhaps best not to assert but merely to suggest an alternate explanation. The case for UFOs as alien spaceships cannot be disproven in every case, but scientists will remain wary of the phenomena cited by the inexperienced. Any true skeptic (a person who neither accepts nor rejects the truth of something purported to be factual) will hold off on the matter of UFOs, but it is not easy to shed that which one wants to believe, and many want to believe that we are visited by aliens from some other planet. For most, perception supersedes reality whenever the two differ.

20

Planets, Gods, and Constellations

The question at once arises whether medieval thinkers really believed that what we now call inanimate objects were sentient and purposive. The answer in general is undoubtedly no. I say "in general," because they attributed life and even intelligence to one privileged class of objects (the stars) which we hold to be inorganic.

C. S. Lewis, *The Discarded Image*

Every society has made order of the sky and its many stars by grouping the stars into constellations. In addition to order and identification, this has the second purpose of placing worthy gods and mortals in the sky where they can be worshiped and honored in perpetuity. Most star groups portray animals: mammals, reptiles, birds, and fish. Some honor people, real or mythical, and very few are named for inanimate objects in any culture. What would we have in the sky if star groups were renamed today? Elvis, maybe the Beatles and Sinatra, and Marilyn as the most worshiped goddess? We would in all likelihood retain the lion and the horse and dogs, and add a cat or two, and a grand totem in the form of an automobile.

We know that the forty-eight primary constellations now designated in Western society originated with people in the Mesopotamian region about 4,000 years ago. They are about the same forty-eight that were passed down through classical times, Hellenic as well as Roman, into the medieval world in Europe and the Middle East and on into our own times.

Between the Middle Ages and modern times the constellations were treated somewhat differently. We have filled every bit of the sky with new ones so that there are now eighty-eight of them. C. S. Lewis cites one very important difference in that the medieval universe, while unimaginably large, was also unambiguously finite. And one unexpected result of this is to make the smallness of the Earth more vividly felt. The ordered medieval cosmos afflicts us with claustrophobia; our universe is romantic, theirs was classical and really quite alien. *Constellation* in medieval language seldom means, as with us, a permanent pattern of stars. It usually means a temporary state of their relative positions.

In China tradition divided the sky into several hundred constellations, each being much smaller than most of ours. What we term an asterism, a subgroup of stars within a larger constellation, the Chinese mythology would likely consider one or even two entire constellations. Our best-known asterism is the Big Dipper, the Plough in Great Britain. It is only a modest part of Ursa Major, the Great Bear, one of the largest constellations of all, with the dipper representing only the hind quarters and tail of this giant animal shape that extends over much of the rest of the northern sky. But the remainder of the sprawling bear has few bright stars and none to rival the magnificent seven that form the dipper.

The next-best-known asterism must surely be the Pleiades, the tiny group of seven sisters in Greek mythology. Only the sharp-eyed among us can see seven; my tired old astigmatic eyes can spot only five (this after cataract surgery). They were considered by some in Ptolemy's time to make up a separate constellation by themselves, but more commonly they are an extension of Taurus, the bull, since they were alleged to have provided certain pleasures for him. After the bull and the bear, we have no well-known asterisms, but to the astronomical taxonomist there are dozens.

How do we know the time and location of the origin of our familiar friends in the sky? We know the time and place because from every latitude on the Earth except the equator, some portion of the sky is perpetually invisible. From northerly midlatitudes, that portion is centered on the south celestial pole and extends up to a northern limit as many degrees from the pole as is our latitude north of the equator. No one would designate constellations in this southerly region because no one could ever see them. We are also aware that the hidden area corresponding to a specific latitude today is not the same as it was centuries in the past. It slowly changes, due to this long-term motion called precession. This precessional motion slowly shifts the locations of the celestial poles in the sky, so that they make a complete circuit once every 25,800 years. Polaris in Ursa Minor, the small bear, is the Pole Star (the nearest bright star to the north celestial pole today) but 4,000 to 5,000 years ago, another star, Thuban, in the constellation Draco, the Dragon, was the Pole Star.

Hence, we need look for the region without constellations and star names from a particular society and find its center, where, presumably, the south pole of the sky was located at the time. This turns out to be about 2000 B.C. for the star groups we know.

The Hellenistic list of forty-eight constellations was favored by Ptolemy, but it doesn't do for the modern world. Noticeable gaps appear between the bright, familiar stars and constellations, and there is that whole region in the deep southern sky that the Alexandrians were aware of but could not see. With the coming of the telescope and the greater need it imposes on us to avoid confusion in the heavens, star catalogs required the addition of more of them. So astronomers of that time (roughly the seventeenth century) added others to the list. Some of the new ones made the final list and others did not. Vulpecula the fox and Monoceros the unicorn made the final cut, joining such familiars as the two bears, Perseus and Hercules, Orion and his two dogs, and the twelve constellations that form the background for the zodiac. The fox and the unicorn filled in two blank spaces in the sky visible from Europe and the United States, and Pavo the peacock, Horologium the clock, and Volans the flying fish are among the ones that helped to cover that uncharted area in the deep southern heavens, once explorers reached the equator and could see them. However, Bufo the toad, Globus Aerostaticus the balloon, and Rangifer the reindeer were not so fortunate; somewhere along the way, they were among the ones that got left off the star charts in favor of others.

Early in the twentieth century it became useful to astronomers to make areas in the sky, rather than star figures alone, of the eighty-eight constellations we have today. Now boundaries delineate each constellation among its neighbors, much as state lines divide up the United States. The boundaries were established by the International Astronomical Union (IAU) around the time of its first meeting in 1922. This is an organization devoted to astronomy and to international cooperation, both being very dear to astronomers of all creeds. Every major nation in the world belongs to the IAU; it was allegedly the first organization to install both China and Taiwan together among its member nations.

The metaphor of state and national boundaries can be extended to ancient times. Their terrestrial equivalent to gaps between and south of the named forty-eight would be the unsurveyed and unsettled regions beyond the cities and towns, described on maps of the time with legends such as "Here there be dragons."

The IAU has the responsibility of keeping order in the sky, and its names for stars, constellations, asteroids, and comets, as well as craters and other surface features on other planets and satellites, are official around the globe. Many hucksters will come along to sell the gullible "the right" to name a star or a comet, and one has every legal right to use that name. But astronomers, amateur and professional alike, will not honor it unless it carries the imprimatur of the IAU, which then brings the name a worldwide credibility. Comets are customarily named after their discoverers, whereas asteroids are named for scientists and the noted among the arts, along with observatories, cities, and occasionally more frivolous objects, but very few statesmen ever make the list. For this reason no Hitler, Stalin, or Vlad the Impaler can be found anywhere in the sky.

✳

OUR SOLAR SYSTEM is 4.6 billion years old. How do we know that so accurately? Many different methods, geological as well as astronomical, have given close agreement on this point. From its birth at that time until around 3.8 billion years ago, the system was not a healthy neighborhood. The reason for it lies in the process of its formation. Not long after the big bang, which we think occurred about 13 billion years ago, the universe was filled with nebulous matter spewed about here and there. Then some billion years later, this nebulosity began to collapse inward on itself, breaking itself up into huge chunks of what we now realize was mostly hydrogen gas. Gas in an unbounded medium tends to do this; it's an inherent property of the stuff. We can see the process on a small scale whenever a cigarette burns quietly in a room in which the air is still. The smoke rises from the lighted tip straight upward for several inches in an even flow called laminar flow. Then since the smoke has no solid boundary, the flow tends to become turbulent; the smoke curls upon itself and breaks up into eddies and whirlpools, most properly called vortices. On a cosmic scale more or less the same process breaks up matter into clouds. In time these clouds will begin to collapse inward upon themselves because they are not at all rigid, and the mutual gravitation of their composite particles pulls them together into ever smaller and denser clouds.

Again, the turbulence that comes about gives the whole cloud a rotation around an axis through the center. At some point the centrifugal force directed outward counterbalances the gravitational collapse, and further collapse is brought to an end, but only in the plane of the rotation. The rest of the material continues to collapse; no outward spin is there to stop it from doing so.

Around 4.6 billion years in the past, most of the hydrogen gas kept right on collapsing, and collapsing gas heats up more and more until another force comes along to balance it again. The cloud or most of it condenses further to the point where the central region reaches a temperature of tens of millions of degrees. It is at this point that thermonuclear fusion takes place with an outward force to balance the inward gravitational contraction and the nebula has become a star. Through either of two complex processes, four hydrogen atoms fuse into a single atom of the second lightest and simplest element, helium. This hydrogen burning, as it is known, is the same process as occurs in a hydrogen bomb. Such a weapon contains a more orthodox uranium or plutonium bomb, which when detonated gets that all-important central temperature to the level that hydrogen fusion can begin. The bomb runs out of fuel right away but the star can take millions or billions of years to do the same. In all this mayhem, the star remains gaseous throughout its entirety.

The gravitational collapse drives the central temperature up to the point where the star could shine, but without fusion it would cool off into a dead object. Astrophysicists know that an original mass must be at least 8 percent as massive as the Sun for fusion to commence. Thus the smallest stars that can shine for so many billions of years have to be that massive at least.

With the start of fusion, the star stops shrinking; here it is entirely gaseous, a furnace made of its own fuel. All this energy gets directed outward away from the stellar interior and out into space. It does so in two ways, known as gas pressure and electron pressure, which together just balance the gravitational pull, and the star falls into equilibrium, staying the same size for most of the rest of its life. The Sun, then, we believe to be in a very stable state for 10 billion years, half of which is time gone by. Having 99.8 percent of all mass in the solar system, it is as massive as a thousand Jupiters and a third of a million Earths. Its diameter is about 110 times that of our planet and its surface temperature is about 5,700°K (11,000°F), and we know that there must be a temperature gradient between the core where it is hottest and the surface where it is coolest.

It might seem that nowhere would physical conditions be more unimaginably impossible to perceive than inside the mighty Sun. We cannot look directly into the Sun or any other star, but since midcentury, astrophysicists, mostly from theory and physical principles, have constructed a sound and consistent picture of the guts of the Sun and other stars. The central solar temperature is about 15 million degrees on the Kelvin scale, similar to that at ground zero in a thermonuclear blast and for the same reason. This we know in part from the core from which subatomic particles called neutrinos are emitted. Neutrinos are spinning massless particles that flow unimpeded outward through the interior of the Sun, and to and beyond the Earth. Here is one of the few ways that stellar interior theory can be checked by observation. With what has to be one of the most unseemly telescopes anywhere, astronomers can count the number of solar neutrino units, or SNUs, that pass through a huge tank filled with about 100,000 gallons of chlorine in the form of cleaning fluid. The tank is located in a gold mine a mile underneath Lead, South Dakota, in the Black Hills (this is a telescope?). The count of SNUs per interval of time is related to the central temperature of the Sun and is thus a measure of it.

Now Jupiter, our largest planet, being only a thousandth the mass of the Sun, is but one eightieth the size needed to heat up its center and become a star. At no time will Jupiter be a star; it has never been one and it will not become one. In the course of the nebula's contraction into the Sun, it left debris orbiting in the plane of rotation. This substance also had its own gravitation and rearranged itself into distinct rings that coalesced into clumps that eventually became the planets.

The four rings near the Sun were affected by its heat and its proximity. The heat warmed them up and the nearby mass gave rise to great tides—and the two processes, tidal and thermal, drove the volatile gaseous hydrogen and helium out into space. What was left was stuff made of the heavier elements. So the condensing blobs, which we could call proto-Mercury, proto-Venus, proto-Earth, and proto-Mars, were formed of the remaining denser elements, the most abundant of which are oxygen, silicon, carbon, iron, and nitrogen. Since together these heavier elements constitute only 3 percent of the total matter of the solar system, the four inner planets ended up small and dense. Their original atmospheres would have been mostly made up of the two lightest elements, hydrogen and helium, but these they were unable to retain. We call this group of inner planets the terrestrial (earthlike) planets.

The planets that formed out in the suburbs of the solar system were far enough from the Sun and cold enough to retain all their original mass. They endured as worlds consisting of mostly hydrogen and helium; they remained large and not very dense, rather like the Sun but without the sunshine. They are called the major or Jovian planets after their largest member, Jupiter. The others are Saturn, Uranus, and Neptune.

Some material did not coalesce into anything. It cooled off into small fragments, most of which lie between Mars and Jupiter. These are the asteroids or minor planets, and the combination of gravitational perturbations of Jupiter and the Sun precluded the formation of a planet there. Out beyond Neptune the same thing happened; matter out there was too sparsely distributed in space to collect into large planets, with the partial exception of Pluto, just large enough that it can be considered a planet or the largest asteroid; only a matter of semantics gives rise to the confusion. No single definite limit separates a planet from an asteroid, any more than one separates a man from a boy. A number of asteroids smaller than Pluto have been recently discovered out near and beyond Pluto, at 40 or more astronomical units from the Sun. But a major difference occurs between this outer belt of asteroids and the ones between Mars and Jupiter that results from their mean distance from the Sun. The latter are close enough to melt or sublimate water from ice to vapor directly, as happens in space, and thus are made of the metals or rocky material that is left behind. The outer bunch are mostly formed of water ice, but with some stony stuff mixed in, and are sometimes referred to as dirty snowballs. Comets are also dirty snowballs since they, too, formed far in the outer limits of the solar system. Sometimes they appear in our nether regions due to gravitational perturbations of Jupiter and the other outer planets upon their orbits.

As noted earlier, throughout the first several hundred million years much debris was left over to bang into the planets and each other, and the system got cleaned out of most of it. One large chunk may have hit a still condensing Earth and knocked a large piece out of it, which became the Moon. Gradually the terrestrial planets developed secondary atmospheres out of the material that shot out of their volcanoes in a process called outgassing. The heavier elements—carbon, oxygen, nitrogen—provided the gas, with carbon and oxygen combining into carbon dioxide. Mercury was just a little too close to the Sun and therefore too hot to keep more than a trace of an atmosphere, and the Moon was just a little too small. But Venus, Earth, and Mars all have considerable atmospheres.

About 3.8 billion years ago, the solar system was emptied out of the big chunks flying around, and on one planet, at least, life was able to form. We are fortunate to have emptied out most of this celestial detritus, but we still have an occasional errant asteroid around; the last large one to collide with our world did so 65 million years ago in a catastrophe that killed off more than half the species alive at the time, including all the dinosaurs. We have only to go back to 1994 to come upon a large object ramming another planet. That year, the comet Shoemaker-Levy 9 got its orbit tangled up with the orbit of Jupiter. After a close flyby past the giant planet, its tidal shear ripped the comet into more than a dozen fragments, each of which returned the favor by ramming it head-on. Each fragment tore into the Jovian atmosphere with force far in excess of the energy that could be mustered from all of our nuclear arsenals detonated at once.

*

LOOK UP AT the sky on a clear night. The firmament looks dark. You can't see anything moving as you watch those serene silent stars. That is why they are still referred to loosely as the fixed stars. Yet after some minutes have elapsed, we realize that they have moved.

The simplest, fastest, and most quickly detected of all motions is the rotation of the stars in the sky about us, reflecting our own planet's rotation on its axis once every 23 hours and 56 minutes. The reason for this value instead of 24 hours as the true day or rotation period was explained earlier, having to do with the other best-known motion, that of the Earth in its orbit rounding the Sun each year.

Think now of the direction you are moving due to each of the two motions. At each instant you are moving toward the east point, the spot on the true horizon due east. Over time, that direction will change, curving downward and to the left in northerly latitudes. The velocity toward this east point is about 0.5 kilometers per second which is the speed unit astronomers use most frequently, or about 1,040 miles per hour at the equator. At other latitudes, the speed is 1,040 miles per hour times the cosine of the latitude, lowering to zero at either Pole.

Our motion in orbit around the Sun is much faster. The orbit is not quite circular, but, as Kepler discovered, it is elliptical by some 1.6 percent. With these slight variations, the average speed comes out to be about 30 kilometers per second or 67,000 miles per hour, more than sixty times as fast as the rotation at the equator. The direction of the orbital motion as one would see it points toward a spot on the ecliptic and about 90° to the west of the Sun. The angle with the Sun varies a bit with the season as we swing in our elliptical orbit from perihelion to aphelion and back again. These are the two motions we know apply to our world since the time of Copernicus. But they are not by a long shot all that we experience. We can trivially add two others by walking along the aisle in a moving railroad car or airplane. But other, astronomical ones lengthen the list as well.

Sir Isaac Newton was the first to realize that the center of gravity, called the barycenter, of two bodies in orbit about each other is not at the center of either, but falls on a line connecting their centers, and closer to the more massive of the two, much like a teeter-totter where the heavier parent must perch closer to its fulcrum than the smaller child for the thing to work. Consider the example of the Earth and the Moon; we know that the larger Earth weighs eighty-one times as much as the Moon. Therefore the center of the larger world will be only $\frac{1}{81}$ of the distance from the barycenter, as is the Moon's center. Since they are separated on average by 240,000 miles, the barycenter lies just under 3,000 miles from the center of the Earth, or 1,000 miles below its surface. We revolve about this point in the same length of time as it takes for the Moon to orbit it once. As noted earlier, the true period of the Moon's revolution is just about $27\frac{1}{3}$ days, and Newton's laws require the Earth to take just that length of time to swing about the same barycenter, which is always located about 1,000 miles directly below the point on its surface that sees the Moon at its zenith. This small motion is negligible in any but the most precise observations of our position within the solar system. Our world pokes along at a speed of just 28 miles per hour, about that of an automobile on a city street. In contrast, the Moon must haul along at eighty-one times this speed, or about 2,268 miles per hour, almost twice the velocity of the Concorde.

What about the Sun? We spin around it and we have mass, not much but enough to move the immense Sun just a tiny bit. But at a third of a million times the Earth's mass, the ponderous Sun is barely nudged along at a sluggish one fifth of a mile per hour, a whole 9 centimeters every second, about the speed of a turtle in a hurry, around a barycenter just 280 miles (450 kilometers) from its own center. Jupiter does better; that most bulky planet pushes the Sun around a point just above its own surface at 12 meters per second or 30 miles per hour. Small as this is, we can detect similar-sized motions of large planets of Jupiter's mass around nearby stars by observing wiggles of this amount in the star's velocity, as is shown in Figure 20.1.

The solar system jerks itself around in a very complex manner. Each planet makes a tired Sun lurch around the common barycenter with it by speeds proportional to size and distance, and the Sun dances around a

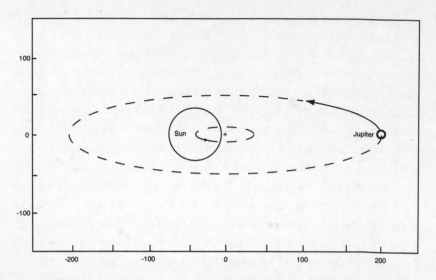

20.1   The Sun and Jupiter orbit about the barycenter (center of gravity) between them. The Sun is about a thousand times as massive as Jupiter; thus it is only a thousandth as far from the barycenter. The Sun moves only about 30 miles per hour in its small orbit, whereas the planet moves a thousand times as fast.

common center of gravity for the system as a whole, the one point that remains stationary with respect to all of its members. The planets are all pulling and hauling on each other, making of their paths round the Sun not the neat ellipses of Kepler, but osculating orbits (here, to osculate means not to kiss but to move in ever-changing, slightly different orbits) around our central star.

Then we have the very slow motion known as the precession of the equinoxes, or simply the precession. Over the eons, this axial wobble of the Earth shifts the constellations mightily in the sky, but like all the other motions listed above this, too, is Earth-centered and thus all of the stars move together in lockstep. Even over its 26,000-year period, the constellations remain undistorted and in their proper positions with respect to each other. No, we must leave the solar system and take to the stars themselves to find their individual motions and, from them, our solar system's progress among them.

21

Our Milky Way and Other Galaxies

In the years after 1543, when Copernicus proposed and published the concept that the Sun and not the Earth lay at the center of the solar system, the nature of the stars first became of individual significance. His heliocentric theory stood on a par with the geocentric universe of Aristotle and Ptolemy for decades until indirect evidence made it the cosmos of choice. Within about 50 years, the English astronomer Thomas Digges and others realized that in the heliocentric arrangement, stars must be seen to exhibit the motion called heliocentric or trigonometric parallax, the inevitable motion that reflects the Earth's orbital motion about the Sun. If that motion cannot be observed, and despite many attempts it wasn't until 1838, the stars must be extremely far away, far beyond Saturn, the most distant of the known planets, making the parallax motion too small to be detected earlier. If that is the case, they cannot be planets, but must be bright like the Sun, and therefore they must be stars as we now know them to be.

For a while, astronomers talked of the stellar system; much larger than the solar system, it embraces the visible stars and more—how much more was uncertain. After Galileo saw that ever fainter stars are visible with ever improved telescopes, further research into the numbers of stars remained rather dormant until Sir William Herschel initiated his studies of their arrangement in space. By training as a musician, a composer with some of his works even now recorded and available, Herschel was the first to explore the cosmos beyond Uranus, which he himself discovered quite by chance in 1781 at his home in Bath, in western England. He named it Georgium Sidus, perhaps to curry favor with his monarch, George III, but the classical tradition took over and the Greco-Roman nomenclature was retained with Uranus, the god of the sky and an Olympian great taking over that remote world.

Then Herschel observed and counted stars, and star counts became a staple of the astronomy of the nineteenth and twentieth centuries. Making order of a jumble of stars with no apparent pattern is a very painstaking business, and in a long lifetime of doing so he established the first meaningful structure to it all. Herschel made counts of stars by magnitude and found that their densities in space decreased in all directions with distance from the solar system. Along the plane of the Milky Way, the decline was gradual, but in the perpendicular directions it was more precipitate. He was led to assume a disklike, grindstone-shaped universe with the Sun near the center; this seemed appropriate since that luminous band was known since the time of Galileo to be composed of millions of stars, most too faint to be seen individually. Toward the directions perpendicular to the plane of the Milky Way, stars, particularly the fainter stars, thinned out rapidly, whereas in and near the plane they fall off only gradually with distance from the Sun. Just as the Earth has a plane (the equator) and two poles, north and south, so does the solar system have its plane, the plane of the planets' orbits and therefore very close to the plane of the ecliptic, also with two poles along the perpendicular to the plane, the north and south poles of the ecliptic or close enough. Not to be outdone, the Galaxy also comes with a plane and two poles. Its angle of inclination to our celestial equator is a whopping 62°, much larger than the tilt of the ecliptic at only 23.5°, as we have noted. In fact, precession will carry the north celestial pole from its present position near Polaris to a point within about 5° of the galactic equator in about 6,000 years. At that time the two equators will be inclined to each other by 85°, almost a right angle.

The Milky Way is like the solar system in many ways, but on a much grander scale. It, too, formed out of some primeval slop to condense under its own gravity until it picked up a rotation. This rotation stopped the condensing as it did in the solar system, in its own plane, but the Galaxy condensed much further, flattening onto a giant disk. We see that disk all around the sky because we are in its midst. But there are differences, big differences. Whereas the solar system stuff condensed into a star and some much less massive planets, the Galaxy did not, because very large blobs of stuff just can't form into single stars—at best a huge blob would make of itself a multitude of stars forming a cluster. Stars can't exist more than at the very most around a hundred times the mass of the Sun, if that. Our galaxy contains about 100 billion times as much matter as is in the solar system; thus the central bulge, corresponding to the Sun's position in the solar system, could condense only into billions of stars, not one or a few superstars.

Since gravity is real and since we are nowhere near the center of our galaxy, we must be in orbit around that center. The Galaxy rotates just as the planets orbit the Sun; here at the Sun we are about 25,000 light-years away from the center and we revolve around it in something like 250 million years. The solar system has thus circled the galactic center about twenty times since its formation. The whole galaxy is older than we are, about three times older; it formed perhaps 12 to 15 billion years ago, not very long after the big bang started it all. With a few hundred billion stars in it, our galaxy is nonetheless only one of billions of galaxies, many of which are much larger than ours. How, talking about this, Carl Sagan could deny that he ever said "billions and billions" is hard to fathom, for astronomical things are truly astronomical in extent.

Above I said that Sir William Herschel determined that the density of stars in space declined in all directions from the Sun. Then how is it that I can also maintain that the Sun is so far from the galactic center? The reason is that stars form only one of the major mass components of the Galaxy; the others consist of interstellar gas and dust, nebulous matter lying between the stars. This is the same substance that formed the solar system so long ago. Does this mean that new stars and solar systems might still be forming out of these nebulae today? The answer is a distinct yes, and we can even see a modern star factory with the naked eye. In the wintertime Orion, the brightest of all constellations and about the best known after the Big Dipper, can be seen to contain a reddish fuzzy spot just south of the three bright stars that form his belt, among three fainter stars that constitute the sword. This is the great Orion nebula, and some stars within it may be only a few hundred or thousand years old. The four bright stars close together that appear to form a small square, a trapezoid really, are together called the trapezoid or trapezium. The four bright stars were born together only a few tens of thousands of years ago out of the swirling eddies of gas lying all around them. More stars are being born there now.

Until early in the twentieth century, scientists continued to believe that the solar system was at the center of the whole system of stars and nebulae, the aggregate of everything we can see. Then some astronomers came to understand that interstellar gas and dust is present between the stars and the visible nebulae; this is the primordial ooze that nebulae are made of. Like the rest of the universe it consists of hydrogen gas, with a little helium mixed in. Then two momentous discoveries were made. Our universe turned out not to be a flattened thing a few thousand light-years across with us at the center, but a galaxy more than 100,000 light-years in diameter, with us out in the suburbs nearly 25,000 or maybe 30,000 light-years from its center. Then to humiliate ourselves further, we found that among the nebulae, those of a spiral form were not nebulae at all but were other galaxies, some of them larger than our own. All these things came together about 1920 and were duly reported on the front page of the *New York Times,* but unlike the Copernican wars of 400 years earlier, it did not turn into a theological dispute as well. Religion had learned to let science work this out by itself within astronomy, if not yet in the field of biology.

How do we move within the swirl of stars about us? Again Sir William Herschel was the first to contemplate the motions of the Sun and the other stars. In time, several years or longer, the nearer stars will be seen to move with respect to their more distant neighbors in the sky. Their angular motions across the celestial sphere are known as proper motions and are measured in seconds of arc per year, or per century if they move slowly. We must then know or assume the star's distance in order to transform its angular proper motion into an absolute linear speed. This Herschel could not do directly, because astronomy wasn't yet up to measuring a valid stellar parallax, from which the distance is found at once. Not until 1838, as mentioned above, did the first parallax get measured, it turns out, by three observers working independently, one of whom was among the first to observe from a southerly latitude, in his case South Africa. The star he chose to observe was Alpha Centauri, a triple star that forms the third brightest star in the heavens, and still the closest to the solar system. Herschel had had to assume that each star shone with the same luminosity as the Sun. This was better than nothing, but not much.

Herschel had noticed that in various parts of the sky, the stars did not move randomly but had a mutual preferred direction. This he correctly assumed was due to the Sun's own particular motion among its neighbors, and the stars reflected this motion by appearing collectively to move the other way. On top of their own scattered individual proper motions, they revealed ours as well. Herschel and many others since have found that the Sun moves, carrying along all the planets and other members of its system, toward a point in the sky in the large sprawling constellation of Hercules, not far from the bright star Vega.

Vega lies in the northern skies of summer and outshines all but one of its fellow stars seen at that time of year. The only brighter blue object anywhere in our skies is Sirius, the most brilliant star of all (after the Sun), located low in the winter sky not far from the point opposite Vega. From the latitudes of Europe and North America, each rises just before the other sets, so we can just see them both near the horizon on opposite sides of the evening sky in the spring and again in the autumn. Our solar retinue is moving more or less toward Vega and away from Sirius at some 20 kilometers (12 miles) per second, or about 4 astronomical units per year. We will not approach Vega anytime soon because it will drift away in another direction well before we get there. It is 27 light-years or 1.7 million astronomical units distant; at our speed it will take nearly half a million years, several ice ages, to get there. Sirius is but 9 light-years off, so it required only a third as long for us to move from its current location. At present, we have launched and placed the two *Voyager* space probes on trajectories that are taking them out of the solar system on galactic orbits of their own. In some part of a million years they may find themselves well away from us and near Sirius.

As one who has done research for years in this field, I know that this solar motion among the other nearby stars that form what we call the solar neighborhood is the very devil to calculate. The velocity, that is, the size and the direction of the solar motion, is highly dependent upon which kinds of stars are used to form an average backward motion assumed to reflect our forward motion on toward Vega. Stars come in a wide range of luminosities; some are tens of thousands of times brighter than our Sun, and others are that much fainter. Furthermore, we know that the fainter the star, the (much) more abundant its kind is in space. For every supergiant Deneb or Rigel in the galaxy, shining like light-houses among fireflies, there are thousands as bright as the Sun, and for every Sun there are dozens of little faint stars. Of the hundreds of stars of the fourth magnitude or brighter, just two are truly (slightly) fainter than the Sun. But conversely, and one is tempted to add diabolically, 90 percent of all stars are fainter. Of the fifty closest stars to us, we can see only about ten without a telescope! This is a colossal imbalance leading to all sorts of large and perverse biases and selection effects in their statistics. Any conclusion based on the few bright stars is as likely to be biased and unrepresentative as would any poll representing the opinions of the rest of us based only on the famous whose names adorn the headlines.

In the study of the kinematics and dynamics of the Galaxy, we want to separate the Sun's own individual motion among its neighbors from their nearly common and much larger orbital motion around the distant galactic center. We and most of the stars we see are blustering along in nearly circular orbits around the central hub of the Galaxy at speeds of some 250 kilometers per second, a dozen times our own solar motion not shared by them. This can be likened to a cluster of gnats flying along. Each gnat is moving with respect to the other members of the swarm, yet all share a motion as they flit from one picnicker to another.

All this leaves us with two ways to define the local standard of rest, or LSR, the name for the point at the Sun that does not move when averaged against our stellar neighbors, or does not move except to describe an exactly circular galactic orbit about the center at our distance from it. The first LSR is called the kinematical LSR, and the other, the dynamical LSR, and they differ by something like 10 kilometers per second. Fraught with errors and uncertainties, this and other findings point to one or, if you like, two additional ways in which the Earth moves among the stars.

✳

RESPLENDENT AS OUR great Milky Way is, it is known today to fill only one tiny corner of the universe. It, too, moves when compared to its billions of neighbors, although the exact nature of that motion is not clear. With the naked eye, we can see three galaxies in the sky external to our Milky Way. Two of these galaxies are relatively small and close to us. These are the Magellanic Clouds, large and small, or LMC and SMC; they were discovered by Magellan and his crew on their voyage around the world in 1521 when they also discovered the Straits of Magellan near the southern extreme of South America. The two clouds lie very close to the south celestial pole and cannot be seen very far north of the equator. They lie about 180,000 light-years off and are therefore not much farther than the width of our galaxy. They are truly satellites of it and orbit about it.

The remaining galaxy is the farthest object visible to the unaided eye; hence its distance answers the question, "How far can you see?" This is Messier 31, M31, a catalog designation, but it is better known as the great galaxy in Andromeda. It is a near twin of ours in size and appearance and lies nearly 2 million light-years from us—the light we see left it at the beginning of the Pleistocene epoch, when humanoids were about as brainy as the great apes. With the large telescopes at Mount Wilson and elsewhere, we could resolve nebulae, star clusters, and supergiant stars that were a match for ours, and thus could calibrate its distance from the difference in brightness between its brightest things and our brightest things. There are now more than twenty galaxies in the region of space between us and the Magellanic Clouds and M31, most of which are puny things by comparison with us and M31, the two giants of this "local group" that we know to be gravitationally bound together, mutually attracting each other.

Then along came Edwin Hubble, an astronomer who observed on the 100-inch reflecting telescope at Mount Wilson near Pasadena, the world's largest from 1918 until the 200-inch telescope at Mount Palomar went into operation in 1948. He and others observed many galaxies farther away than M31, estimated their distances, and from their spectra he concluded that they are receding from us in direct proportion to their distances. From the Doppler effect, we know that light waves, much as sound waves, are compressed toward bluer light from objects approaching the observer and extended toward the red from those receding from us. Galaxies beyond the local group are all moving away, and the farther the galaxy, the faster the recession.

Hubble had discovered that the whole universe is expanding, growing larger as time goes by. From each point within it one gets the identical impression that all galaxies are receding in proportion to their distance from oneself. Here is the last and perhaps fastest way in which we are hurtling through space, but with respect to what? Einstein and others showed that there is no fundamental center of the universe. Alternatively you can place the center wherever you want—in downtown Boston or at the South Pole or on Jupiter's largest moon, wherever.

The Hubble expansion rate (the Hubble constant, H) is thought now to be about 70 kilometers per second per megaparsec (mpc, 1 million parsecs or 3.26 million light-years). Before about 1950, the Hubble expansion was held to be some 500, but if extrapolated backward in time, everything was together only a couple of billion years ago, an age younger than the Earth. That just won't do, and a recalibration of distances to galaxies was in order and was soon made. A much lower value of H was necessary so that galaxies took longer to get out to where we believe they are. After much haggling, astronomers are agreed, more or less, that H is near 70. This means that everything was together about 13 billion years ago. At that time, the big bang took place and the universe blew apart, a view supported by much in the field of physics. With each larger telescope and superior auxiliary detecting equipment, we see galaxies with greater red shifts and therefore with greater recessional velocities. At present the farthest objects lie billions of light-years from us and are moving at about 94 percent of the speed of light. The observed universe just keeps getting bigger and bigger and no one knows if and where there is a limit to the whole thing. If we knew the average density of matter within the universe, we would know whether its mutual gravitation is strong enough to pull it all back together again, presumably for another big bang to occur, or too weak to do that, in which case the universe will go on expanding forever. Observations cannot now make a distinction between these two very different models for the future universe. Stay tuned.

22

✳

Planets, Stars, and Drake's Equation

Frank Drake was among the earliest astronomers to become seriously interested in the detection of alien civilizations. As early as 1960, we earthlings had the capability of detecting another intelligence of just our own level of technology, should one exist among the nearer stars. At that time, detection was limited to radio wavelengths as the new and rapidly expanding field of radio astronomy came to its maturity. Drake and others launched Project Ozma, the first serious attempt to contact extraterrestrial intelligence. Two nearby stars, both only about 11 light-years away, were the first to be listened to for any signals indicating intelligence from a nearby planet. The stars are known as Tau Ceti and Epsilon Eridani and are of magnitudes 3.5 and 3.7, just nicely visible to the naked eye from the suburbs, and they are, as it happens, the two brightest stars in the sky that are intrinsically fainter than the Sun. Such a bias exists among luminosities of the stars that all of the three-hundred-odd stars brighter than these are intrinsically more luminous than the Sun, despite the reality that about nine stars in ten in our galaxy are fainter. Think of the overwhelming predominance of the truly bright stars this way; if all the stars brighter than the Sun were to disappear, our eyes would spot these two stars and about half a dozen others of the fourth and fifth magnitudes, and that's all! From southerly latitudes a single first-magnitude star would remain, because the fainter of the two bright stars that appear as one to the naked eye is a bit fainter than the Sun. Being the closest at only 4 light-years, this star would appear alone at magnitude 1.4. Ninety percent of all stars would still be there, but they are just too faint to see.

Drake and his colleagues heard nothing, and despite enormous strides in detection capability in the following four decades, no signal has yet been recognized as of a design of intelligence. The organization known as SETI (the Search for Extraterrestrial Intelligence) is today a NASA-sponsored search program of enormous range and sensitivity; perhaps he and others will succeed soon in their quest.

The problem of estimating the chances of hearing any neighbors in space is huge and intractable. Drake modeled it as best one could in an equation still known as Drake's equation. It looks like this:

$$N = Rf(p)n(e)f(l)f(i)f(c)L$$

where N is the number of civilizations in our galaxy able to contact each other. N is the product of seven different factors, none of them known well, if at all. The first is R and stands for the rate at which stars form in the Milky Way, and the second, $f(p)$, stands for the fraction of these stars that have planets. The next term, $n(e)$, is the number of planets per solar system suitable for the existence of life, and $f(l)$ represents the fraction of the planets upon which life actually arises. The term $f(i)$ is the fraction of these life systems that develop intelligence necessary for a civilization like ours, and $f(c)$ is the fraction of intelligent societies that choose to develop technology and use it to communicate with others. Finally, L represents the lifetime of such a society. L would not have been very long for us if we had blown ourselves to pieces during the Cold War, but we could last for millions of years if we behave ourselves and deter any errant comet or asteroid from coming our way.

The four terms involving the letter f vary anywhere from zero to unity. We have no idea of the actual probability of any of the last three, and $f(c)$ depends heavily on the proper value of L. Calculating N, then, is purely a guessing game, not even good enough for government work.

Astrophysicists have had some knowledge of the first term, R, but the big news of the day is that almost overnight we can estimate the second term, f(p). Just about 5 years ago, in 1995, the first proof of the existence of a planet orbiting a star other than the Sun was confirmed. Shortly therafter, I wrote (in *Night Has a Thousand Eyes*) of the new discovery of planets in other stellar systems. Now the newer, larger telescopes and especially the greatly increased sensitivities of ancillary detecting devices, along with many more observations, have provided an enormous breakthrough in this field. We have confirmed the existence of dozens of planets elsewhere, and can even give a sensible guesstimate of the size of that second term. Something like 6 percent of all stars rather like the Sun appear to be accompanied by Jupiters, as they are called. These are planets of the general size and mass (and presumably the constitution) of Jupiter, our largest planet, some 318 times as massive as Earth.

Based on our one known example plus other factors, Earth-sized planets are far more likely to spawn life than the more massive Jupiters, but the latter are the planets big enough to detect at this time. When we consider the solar system objects most conducive to settlement and colonization, we center on three: the Earth, the Moon, and Mars. But as we detect planets elsewhere, our focus changes to another triad, the Sun, the Earth, and Jupiter. Since we still do not directly see any extrasolar planet, we detect them gravitationally, and these are the largest. As far as we know, they resemble our largest world, Jupiter. But other Earthlike planets are of peak interest, since our predominant goal is the detection of life and, above all, intelligent life. These must await another turn of the screw toward ever-increasing sophistication in detection, but the achievement may come about in a matter of years, not decades.

If we range upward in mass from Jupiter, we find a plethora of new objects that are several times its mass, yet seem to be and behave like planets. The Sun, a rather average star, is 1,047 times as massive as Jupiter; this thousand-to-one ratio has been daunting to astronomers who wish to know and understand those objects that lie in between the masses of our star and our biggest nonstar. Stellar masses are not easy to come by, despite their enormous importance in any understanding of stellar structure and evolution. From certain double stars, observers have derived masses of the two components, and find that stars seem to peter out above around fifty times the solar mass on the high end to one tenth on the low side (we set the mass of the Sun at 1.00 solar mass by definition in order to avoid awkward numbers like two times 10 to the twenty-seventh power, or 2,000,000,000,000,000,000,000,000,000, the weight of the Sun in tons). The latter stars are extremely faint and it takes a large telescope to render the observations from which masses can be found. Keep in mind that one tenth of a solar mass is still one hundred times the mass of Jupiter.

Some time ago, as the interiors of normal stars began to be well understood, it became clear to astrophysicists that stars have a lower limit to mass, or they are not proper stars. The Sun and the other similar stars can churn their fuel for a long time; the Sun has 5 billion years of its 10-billion-year lifetime left to live. That it does so is due to a gradient in its temperature ranging from about 6,000°K at its surface to 15 million degrees at the center. That is hot enough for fusion to begin, in which hydrogen is converted into helium by subatomic processes that resemble a detonated hydrogen bomb that achieves such a temperature for an instant in the center of its explosion. But a star must have 0.08 solar masses, eighty Jupiters, to achieve this hotness from the heat produced when it collapsed from a nebula into the star it becomes. Smaller "stars" shine for a while, but then just cool off into dark bodies of some kind. After many years of looking, astronomers using the largest telescopes found a few of these almost-stars, those weighing less than eighty times Jupiter. The younger among them appear as red dwarf stars for a while like their bigger brothers, but then just fade away as surely as do old soldiers. *Brown dwarfs* is the term used for them so that they do not get confused with the real red dwarfs that shine for an even longer period than does the Sun.

In the last few years we went from a recognition of none of these semi-stars to several dozen today. Some may be as small as twenty Jupiters in mass. And since we now know some planets orbiting other stars with masses from less than one to around five Jupiters, astronomers are close to the realization of an age-old dream: to bridge the gap between stars and planets. We know little of the physics of the brown dwarfs or the oversized Jupiters, but soon the continuum of masses may be complete and the difference between stars and planets will be understood at last.

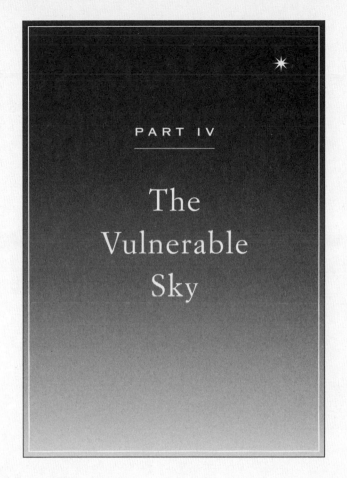

PART IV

The
Vulnerable
Sky

23

The Fate of the Sky

Now, my suspicion is that the universe is not only queerer than we suppose, but queerer than we can suppose. . . . I suspect that there are more things in heaven and earth than are dreamed of, in any philosophy.

J. B. S. Haldane

The Burgess Shale is a region of the Rocky Mountains of British Columbia. It is remarkable because of the fossils and the soft-bodied preservation of a wide variety of invertebrate animals found there. The rock dates from the middle Cambrian age, which occurred about 550 million years ago when this rock was deposited. Just before that age, the single-cell life-forms that had been the only living things in existence for over 3 billion years burst forth in a revolutionary eruption of larger creatures—animals of our size or close. This initial diversification including most of the various animal phyla, or major body schemes, originated within this relatively brief span of time; in fact, more phyla were formed then than are extant today. No one knows what brought about this Cambrian explosion, as it is called, but it started the chain of life that evolved into the dinosaurs and, ultimately, us. No skeletal remains exist from as long ago as 600 million years; thus its suddenness makes *explosion* an appropriate word. Mars may have given rise to billions of years of one-celled creatures, but we know of no other locale that sprouted larger forms as did the Earth in Cambrian times. It is this event that is more likely

unique or nearly so in the whole Galaxy than the origin of life itself; simpler life appears more likely to occur on the red planet and elsewhere. In
all the time since the Cambrian explosion, the rotation of the Earth has
been much as it is today. The day was shorter, as has been explained earlier, but only by a few hours. Almost all life on this planet has experienced about the same circadian rhythm as we do now—alternating light
and darkness forms a pattern familiar to all species. Only the creatures
that live deep in the oceans have not because theirs is a world of total
darkness, a world where the Sun does not shine.

Yet in all that time, only our species, Homo sapiens, has amassed
power and energy for our use far beyond all of the other species that
have trod, swum, or flown on this planet. Whenever we consider our
species as a whole, we defer to God or gods or to Mother Nature. Yet
whenever we divide ourselves into tribes, now called nations, we just as
rapidly fiercely proclaim our might with bellicose fervor. One would
think that the United States (or any nation) alone can muster more
power than the totality of the human race. For a tiny fraction of the time
that the Earth and the other planets have been circling the Sun, we have
prevailed in rearranging much of our world, and yet how quick we are to
prostrate ourselves before natural events when we claim that a single
hurricane or a single volcano emits more energy than our entire nuclear
arsenal and all other energy sources available to us.

Our Earth and its fellow planets have been faithfully orbiting the Sun
for 4.6 billion years—could anything be more permanent? Well, yes and
no; the entire visible universe is now thought to be about three times as
old, some 13 to 15 billion years, since we speculate that it was all in a very
small volume with an incredibly high mass density prior to the big bang.
This big bang blew it all apart and it has been expanding ever since.
Now, billions of year later, it is still expanding, as Edwin Hubble first discovered. Will it continue to expand forever, or will it condense back
upon itself into another blob, only to form another big bang, in a continuing series of cataclysms? We don't know; even those who think they
know probably don't. The problem lies in the amount of matter in and
between the galaxies whose gravitation acts to pull the whole lot back

together again and again. This matter, not likely to be in the form of visible stars but rather in nebular form, cannot be seen or detected sufficiently well to estimate its gravitational potential. It may be enough to close up the universe again or it may not. Observations with the largest telescopes place this estimate right on the cusp between an open universe expanding forever and a closed model forcing recurrent big bangs every several tens of billions of years.

If the mass cannot reconstitute the cosmos, the constituent stars and galaxies will fly farther and farther apart. Since stars coalesce out of the interstellar material, they usurp some of this mass into themselves, and most stars do not give it up again into the void. Some do, but they are a minority, which, through supernova explosions, litter space with second- and third-generation stuff from which new stars can be formed. But in the long run, more and ever more matter gets locked in the form of dead stars and the interstellar medium fails to be continually enriched with fresh material.

What happens when matter gets imprisoned in tight configurations called stars, living and dead? Planets will eventually get detached from stars, stars will become detached from galaxies, and black holes will accrete everything else. The universe becomes a dark black graveyard of civilizations and anything else made of atoms. Some black holes will persist for a while, but recall that the volume of occupied space continues to expand, and the mass does not. The inevitable conclusion is that the density of everything diminishes beyond the point where stars and galaxies and black holes can form.

Recent evidence points toward an outward acceleration of galaxies away from each other; a kind of antigravity propelling them apart at an ever-faster rate. Should this effect be substantiated in full, we would have a third kind of cosmos driven by a hitherto unknown force to consider. This makes for a new ball game and might lead back to the supposition that ours is only one of many universes, collectively making up a sort of "multiverse." That would help those who feel that the works of Shakespeare, Michelangelo, and Beethoven are truly immortal and shall, as is proper, never perish from some corner of existence.

*

THE SUN, MOON, and planets circle through the heavens in their predictable ways against the background of the so-called fixed stars. But over the ages, these fixed stars move, too; they subscribe to the great year of the precession, looping about up and down, north and south, over a cycle of 26,000 years. Furthermore, each star has its own proper motion that swings it past the solar system and then into the background of fainter, more distant stars. These stars also do the same thing over a longer period of time. Would not this rhythm repeat itself endlessly?

The firmament visible to the naked eye consists of these individual stars; only the great nebula in Orion, the great galaxy in Andromeda, and a handful of other things appear diffuse to us, and none are very conspicuous. Even the nebulosity of the Milky Way itself is nothing more than several hundred billions of stars, as Galileo was the first to observe. Each star is too faint to be resolved, but collectively they cast a ring of faint light around the celestial sphere. But the diffuse objects hold the story line into eternity. Throughout the lifetime of the solar system the stars may appear to dominate, but over the billions and billions of years, the nebulae and galaxies show the slowest changes of all. In time the other galaxies will either rush in upon us from all directions in a stampede of light toward the next big bang, or they will all disappear forever. If the latter, the universe will die the death of corpses of stars with no light forever roaming an ever-greater volume of space. In either event, the human race will long since have disappeared along with all other life. But long before that interminably distant epoch, we can keep the sky we have.

24

Light Pollution:
A Change of Paradigm

There is the greatest difference between presuming an opinion to be
true because, with every opportunity for contesting it, it has not
been refuted, and assuming its truth for the purpose of not permitting its refutation.

John Stuart Mill, *On Liberty*

Read not to contradict and confute; nor to believe and take for
granted; nor to find talk and discourse; but to weigh and consider.

Francis Bacon, *Of Studies*

For years I have been asked for advice on how to reduce bad lighting on
college campuses. I have some experience, but not much in the political
or Machiavellian realm to offer, and no magic solutions. In the 1960s and
early '70s, my first years of teaching astronomy at Wesleyan, I found an
ever brighter campus and my complaints at all the increased excessive
lighting fell on deaf ears. I was the chair of the astronomy department for
years but without tenure at first; therefore I had to tread lightly if this
department was to survive. Horatius at the bridge I was not; I had no
reserve to withstand the onslaught of lumens for perceived safety, especially after women were admitted to Wesleyan around 1970. With a pal-
pable sneer, students and faculty alike suggested that I take a rifle to the
lights if they offended me. "More light is always better" and "lighting

deters and defeats crime" on and off the campus were the reigning paradigms of the time. The responsibility for campus lighting was handed over to the Student Affairs Committee, which continued to refuse to hear my case, until the dean of students required them to do so in accordance with university rules. Regardless of the level of skill at campus politics, my outstanding negative lay in the reality that I was alone, a single voice. Curiously some of my colleagues of my own astronomy department later became as hostile as anyone. They fully appreciated a dark sky, but felt I was rocking the boat, and they might suffer the fiscal consequences in future department budgets (a fear that did not come to pass). I was as much a pariah there as elsewhere, and was left in the dark in decisions made about lighting.

The ever brightening night sky had the effect of denying us astronomers access to our laboratory, and it threatened the continuation of two federally funded research programs in which students participated. For years as chair of an astronomy department and director of a research observatory, I had to contend with the loss of the sky to my students and faculty alike. Scant interest and not a little hostility from the university community were shown in shielding campus lights and otherwise reducing sky glare until later. Imagine the uproar if the physicists or chemists had been denied access to their labs! But, you see (it was carefully explained to us), we need more lights to drive the criminal element away from the campus.

In a rather halting manner, I made my case for the constraint and control of outdoor lighting. I was about to put to the test John Stuart Mill's adage that "one person plus the truth make a majority." I was more inclined to accept as reality a remark the maverick Washington journalist I. F. Stone made in my presence, that "one advocate is an evangelist, whereas two or more make a pressure group." In those not so bygone days there were no shielded fixtures, no studies questioning the overuse of night lighting, but some advocacy of lighting to saturation and beyond by utility companies and others with a financial interest. What would I have the campus planners do? This was a valid question at the time. Removing even the most offensive lights was not in the cards on any campus in those days.

I don't recall now which of the ways I took to educate the educated at Wesleyan about the dark side of outdoor lighting were of my own invention, and which were borrowed. Perhaps the first success came after someone installed four floodlights, each rated at 100,000 lumens, atop four buildings that surrounded Andrus Field in the midst of the campus that shone all night, every night. Two of them were directed right toward the dome of the 20-inch telescope of our Van Vleck Observatory, each being seen there at about ten times the brightness of the Full Moon. At my instigation, I met at night on Andrus Field with the dean of science and members of Public Safety and the Physical Plant. I was the last to arrive and came upon the group from the direction of one of the lights. Were I an armed assailant, not one of them could have taken defensive action until too late, and they knew it. This was an example of direct glare, never a useful or desirable lighting feature. The four lights left soon afterward.

On the cover of the annual department report to the trustees, I placed sketches of shielded and unshielded light fixtures. Good light need only shine down; not even Speedy Gonzales could harass a student above a lamppost, where light is truly a waste in energy and money. Finally I obtained permission to shield a few lamps nearest the observatory. Some student majors assisted in taping the glass sides of the mansard-style luminaires, but when one student on a ladder was almost pulled down by other students, I continued by myself. I also met with a committee of faculty and trustees to plead my case, and brought to their attention a recent Justice Department study that concluded that "while there is no statistically significant evidence that streetlighting impacts the level of crime, especially if crime displacement is taken into account, there is a strong indication that increased lighting—perhaps lighting uniformity—decreases the fear of crime." Since 1979, when this study was published, supportive evidence has come in from many quarters that outdoor lighting does not deter crime. It may speciously abate our fear of crime, which would lead to misplaced trust and a false sense of security, and the danger that real preventives may go wanting for funding and action.

Then, sometime during these developments, onto the scene like the Lone Ranger on his horse (white, of course) came a bunch of merrie men gathered into a new group called the International Dark-Sky Association (IDA), and I joined it on the very day I first heard about it. Here was the badly needed nucleus of a pressure group, and some people who knew more, not less, than I did about excessive lighting and from whom I could learn. The IDA, founded in Tucson, Arizona, in 1988, started small with a membership mostly of astronomers observing with one or another of the large mountaintop telescopes that surround the city. I participated in its first conference, held in Washington in August in that same year, 1988. The astonishing growth and success of the IDA since that time are due in part to its insistence on positive cooperation and discourse, eschewing negative confrontation when dealing with lighting advocates. More than any other organization, it brought to an end the time, as Dr. D. A. Schreuder has stated, when "astronomers regarded lighting engineers as lumen-happy peddlers who only lived to sell lamps, and that lighting engineers considered astronomers as egg-heads who did nothing but squander taxpayers' money on their hobby." Practitioners of both professions realize that "we all live on a small and very full planet and that we should respect each other as fellow human beings. Technical and scientific problems are just that; they can be easily solved when we all collaborate."

Hearing the benefits of good lighting from many sources in addition to a lone idiosyncratic professor proved to be the key to Wesleyan's revision of outdoor lighting in 1991 on a plan that I helped to create. Shielding was the major difference, full-cutoff shielding on lamps spaced with sensible restraint. At that time, I knew of no larger area in central Connecticut illuminated only with shielded lighting, although now there are many. The IDA provided the essential nucleus for all of us night watchmen as we watched the night disappear. Working with a cooperative IESNA (Illuminating Engineering Society of North America), we have reevaluated (downward in most cases) the necessary light for a number of nocturnal activities. Today with more than seven thousand members in every state and seventy other countries, its astronomers are outnumbered by lighting engineers and environmentalists among the membership.

The IDA has discovered those who seek to darken our sky to aid turtles and migratory birds confused by light flooding upward, and now joins with them to press the case. For every success, however, a new gas station or eatery erects lighting of overkill proportions and streetlights are added to a new suburban housing development. In Middletown, Connecticut, the local utility company has sought to place about 160 floodlights on a bridge to bathe its superstructure in light and make of it, perforce, a tourist attraction. In this ill-conceived plan, only 5 percent of the million lumens would illuminate any part of the span; the rest would saturate the sky with eternal light and kill or drive off the many migrating songbirds, one of which, the robin, is the Connecticut state bird.

Such mistaken ventures only multiply with each success. But we keepers of the night have successes of our own. A similar lighting plan proposed for Britain's longest suspension bridge, which crosses the Humber River near its mouth on the North Sea, was scuttled in the face of mounting criticism, and another in Los Angeles met the same fate. Many new ordinances at the local and state levels now prohibit the worst in bright, unshielded light. The oldest and most treasured site for amateur astronomers to gather at night is threatened by a prison to be erected only 3 miles away. This is Stellafane in rural Vermont, now a national historic site. Its pollution by light would be a visible travesty to a world seeking relief from pollution, particularly if the prison promotes secondary business and security lighting of poor quality. However, the prison's construction is now planned with a minimum of outdoor lighting.

When asked for advice on coping with an overbright college campus, I can provide mostly generalities, but I feel they are useful. One is a study of my own that I undertook beginning in 1975 and continued through 1990 when outdoor lighting exploded at Wesleyan. During that 15-year period, I visited at night every outdoor site at which the Office of Public Safety stated that an occasion of molestation or worse had occurred in hours of darkness. These instances were compared to the numbers reported as happening in the daytime or indoors or at well-illuminated places at night. The data are given at the end of Appendix I and they confirm the refutation by the Justice Department study and others that dark conditions do not lead to an excess of these kinds of events, most of which involved harassment or molestation of coeds and others on and near the campus. This crime problem has rightfully occupied the central place in efforts to assure safety and security for all as much as possible. But my results imply that increased lighting is not a necessary component of a successful safety program. Furthermore, during the 15 years of my study, only four letters were published on the subject of outdoor lighting in the campus newspaper, the *Wesleyan Argus,* indicating that lighting was neither prominent nor widespread among the concerns of most students. Had my case not been widely accepted, the trustees of Wesleyan University would not and should not ever have accepted shielding and the reduction of direct glare on its campus. The paradigm stating that more light is always better became divorced from real concerns for campus security and was replaced, for some at least, by another invoking good lighting as a replacement for bad lighting.

My recommendations for other colleges and universities follow:

First, become a member of the International Dark-Sky Association (3225 North First Avenue, Tucson, AZ 85719, phone 520-293-3198, web: www.darksky.org). The IDA is by far the most knowledgeable and experienced group on the planet for problems of glare and lighting overkill, and the dues are very modest.

Second, obtain copies of the IDA information sheets, the best and most current library source on light pollution available. Make a set of the sheets available to a campus or local library. Invest in some IDA slides and give an illustrated talk with examples of good and bad lighting, or secure a guest speaker to do the same.

Third, invite administrators to an evening on campus with a telescope, and let them see the celestial glories against the lighting glare for themselves. Remind them of the waste in money and energy in upward light and the danger to migratory birds. The astronomical problem should be self-evident, and the bottom line is always of interest to an administrator.

Fourth, tap into the network of dark-sky advocates, if only to become aware that you are by no means alone. Many messages, ranging from the ridiculous to the sublime, reveal a widespread hearty interest in dark skies as a way of life. Try to obtain the design for any specific lighting proposal. Discuss it with the local utility, and the local government. Find out if taxpayers are to share the cost, which may be substantial for both installation and maintenance. Work the word *taxpayer* into discussions with municipal officials whenever appropriate.

Finally, resist any temptation toward confrontation and contentious remarks. A cooperative attitude makes the win-win properties of curbing light pollution much more readily identified as such.

✳

LIGHT POLLUTION IS rapidly becoming a household term. Lately, astronomers, environmentalists, lighting engineers, and those who just yearn for a dark sky have organized in order to educate the public and reverse the ever-increasing night sky brightness. The International Dark-Sky Association is the primary vehicle for the coordination of these efforts.

In the last few years, the IDA has been joined by the Illuminating Engineering Society of North America (IESNA or IES), astronomical organizations such as the International Astronomical Union, and, in 1999, the United Nations. Individual, life, and corporate memberships in IDA have all doubled in the last year or two, and the Sky Publishing Corporation, publisher of *Sky & Telescope* magazine, and several other corporations have given sizable cash awards.

The movement to bring light pollution under control has come into publicity and success as never before. Laws limiting and controlling light pollution, direct glare, and light trespass (unwanted light crossing property lines) have been passed by Maine, Texas, and New Mexico, and many municipalities and counties around the country and the world. It is no longer permissible to erect a light fixture unshielded from above in many of these places. Billboards should be lit downward from above (not upward from below with excess light spilling into the sky) and lighting used in outdoor advertising turned off by an appropriate hour in the late evening. Light trespass (unwanted glare splashing onto the property of others) is now illegal in many places. At present the IDA and IESNA are collaborating in setting upper limits to night lighting for driving, sports, and other nighttime activities to be used as guidelines in further legislation.

As with many widely held beliefs among the populace, the concept that more light is always better pits perception against reality. Reduction of crime from brighter lights or glare is not substantiated by any credible evidence, certainly not if the light leads to sideways glare and upward-directed glow. To avoid these problems, full-cutoff (FCO) shielding is necessary. Full-cutoff is a technical term, defined below.

Lighting manufacturers now make fixtures for streetlights and private lights that conform to FCO shielding, and cost no more than the glare-producing unshielded fixtures found everywhere. Shielding qualifying for the FCO designation is probably the best single adjustment localities can make to reduce wasteful upward lighting and thus lower the brightness of the night sky. This helps astronomers of all sorts, from star gazers to professionals, as well as bird migration and (along seacoasts) sea turtles, whose habitats depend on the normal circadian rhythm of light and natural darkness, of day and night.

Fixtures equipped with full-cutoff shielding come in all styles from American colonial to the common cobra-head lights now prevalent all over the world. They are a bit more complex than might first appear. If we consider θ as the angle between the vertical aimed downward and the direction in question, properly designed full-cutoff luminaires use reflective surfaces to increase the brightness of the light source with increase in θ, from a direction of 0° (straight down) to about 65°, in order to offset the normal diminution of light with θ that occurs because as the angle increases, the point where the light strikes the surface recedes from the pole and the light becomes weaker. The result is a nicely uniform light distribution on the ground to about three times the height of the pole, as appears in Figure 24.1. This avoids a bright spot right under the pole, and dark spaces between it and its neighboring poles are minimized. At a sensible spacing between poles of six times their height, the light along the roadway is nearly uniform. Contrast this with the unshielded light clutter shown in Figure 24.2.

Above the angle θ of about 65°, the light output becomes sharply reduced so that by about 80°, it is close to zero. This reduction combats direct glare, a feature that gives rise to one of the truly useless and dangerous of all driving conditions. Anyone who has driven into a bright sunset

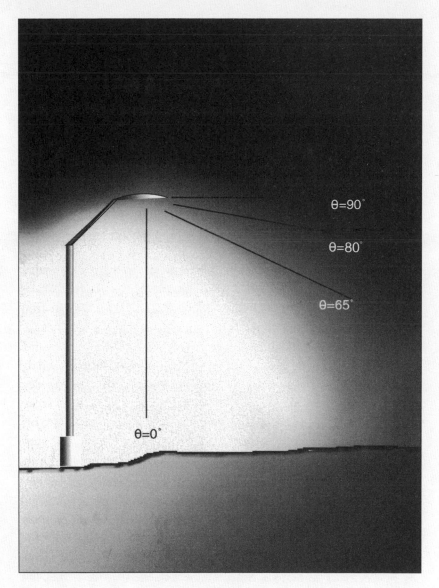

θ=90°

θ=80°

θ=65°

θ=0°

24.1　Diagram showing the brightness of light output with angle from the vertical pointing downward

knows glare. When the Sun is low, when it is within about 10° of the horizon, we recognize it as a great nuisance to the eyes and a hazard to road safety. When the Sun is farther aloft in the sky, say 25° in altitude, it shines with much less direct glare and can usually be blocked by a visor. And after the Sun sets, great relief is appreciated by everyone. The shielded light emulates these conditions. At a 65° angle, the light appears well above the roadway, but at 80° it is low and shines directly into the eyes.

Above the horizon, at angles over 90°, a properly shielded light source or bulb is not directly visible at all and provides no direct illumination of the sky and the natural and man-made aerosols in the air that reflect the light and thus brighten the sky. This leads to the greatest of all benefits of shielded light: the reduction of the waste of energy and money. Light shining only downward can be reduced by about one third in wattage with no loss to safety. The two sketches in Figure 24.3 show the directed output for a shielded and an unshielded fixture. In the latter case, some 35 percent of the total shines above the horizontal plane directly into the sky to illuminate the universe, which does a very good job of that on its own without our help.

The remaining 65 percent falls on the ground, which has a typical albedo (the ratio of reflected light to incident light) of around 15 percent, higher in cases with snow cover. The 15 percent of 65 percent accounts for about 10 percent of the total light. For an unshielded lamp, this reflected light can be added to the 35 percent shining upward, to total 45 percent. For the shielded case only that 10 percent illuminates the sky, which is only one fourth as bright as a direct consequence. Furthermore, the electric consumption is reduced by one third, since only the downward-shining light is needed, and the wattage of the bulb can be reduced by one third without loss of light where it is needed. Most electric power is derived from fossil fuel (coal, oil, gas); hence any lowering of electricity leads to a concomitant reduction in fossil fuel use, dependence on foreign oil, and global warming through reduction of carbon dioxide in the atmosphere. Everyone wins! Were all streetlights in the United States converted to properly shielded lamps, the energy usage could be cut by nearly $2 billion per year and each community would share in this windfall after a cost recovery period of about 3 years

24.2   An example of light clutter. No attempt is made to design the light either for conservation of energy or for attractiveness. Courtesy of International Dark-Sky Association

for the transformation. Since no loss of individual or vehicular safety occurs, this is truly a win-win situation, not a zero-sum game.

A second goal of light conservation, after full-cutoff shielding, is the use of no more light than is necessary. The IES, along with the IDA, is making new recommendations for the legislation of light-level limits in specific circumstances. For example, along streets and sidewalks, these organizations have followed the lead of the National Safety Council, and mandated a level on the surface of 1 to 2 foot-candles (a foot-candle is roughly the brightness of a candle as seen from a distance of 1 foot). This is sufficient to see curbs and potholes and other potential obstacles, but many streets and walkways are many times brighter. In parking lots and sports sites, a greater light level is recommended, as for gas stations and roadside restaurants, perhaps the greatest offenders at present. Certainly 10 to 20 foot-candles suffice to illuminate almost any nighttime activity. But gas stations typically register light intensities of more than 100 foot-candles at the pumps. So bright are these installations that

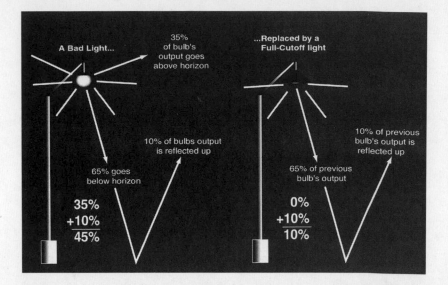

24.3   An FCO lamp saves energy and diminishes sky brightness. The light distribution of an unshielded and a shielded streetlight. The skyglow can be reduced by switching to full-cutoff light fixtures. With reduced wattage the downward light remains the same, but the total upward light is reduced by about three quarters. Copyright *Sky & Telescope* 1996

along thoroughfares motorists can forget to turn their headlights back on when they head for the highway after leaving the gas station. In all of this, it is worth recalling that the Full Moon, at its very brightest, produces only one fiftieth (2 percent) of a foot-candle. Certainly parking lots and other public places need not exceed the Full Moon by more than a factor of 500, as many do now.

25

＊

Glare

Much of what we currently know about what is or isn't so will surely change in subsequent years. What is most important, then, is not dispelling particular erroneous beliefs (although there is surely some merit in that), but in creating an understanding of how we form erroneous beliefs. To truly appreciate the complexities of the world and the intricacies of human experience, it is essential to understand how we can be misled by the apparent evidence of everyday experience. This, in turn, requires that we think clearly about our experience, question our assumptions, and challenge what we think we know.

Three associated problems arose in concert with the growing orange malaise of high-pressure sodium street and other lights throughout the world: these are light pollution, light trespass, and glare. The first caused the near disappearance of the stars above the aerosols, the particulate matter suspended in the air, awash with light photons streaming upward from below. This is known as skyglow, excessive skyglow. The second is the spilling of light onto places and properties where it is not wanted; this is known as light trespass. These problems are joined by an awareness of the third complication, glare.

Glare, direct glare, is a particularly pernicious matter. Only now is it widely recognized as a perilous one, despite the example of years of arduous driving into blinding sunsets. Improperly shielded streetlights can produce glare to the side as well as upward light, creating dangerous conditions for pedestrians and traffic. An assailant approaching from the direction of a bright light may not be seen by the victim in time to take evasive action or to identify the would-be criminal or his clothing. Motorists blinded by direct glare from streetlights shining into their eyes may not detect a pedestrian or animal or another vehicle approaching from the side, particularly under misty or rainy weather conditions. Far more safe and efficient are downward-shining lamps with their light shielded from the side as well as from above. Especially do older drivers find it difficult to spot at night those in the road ahead of them in time to avoid an accident.

Direct glare is defined as the visual discomfort resulting from insufficiently shielded light sources in the field of view. One should be able to see the effect, not the light source. Use of the term *direct glare* is recommended in lieu of the word *glare* alone. The definition of direct glare implies that if you can directly see an unshielded lamp, or the maximum intensity zone of a light fixture, you likely have a glare condition. Whenever one is near such a light source, one will see these parameters to some extent. For this reason, a reasonable definition involves a limit placed on the definition of the field of view. To be sure, instances occur where upward-shining light may be justified, as, for example, the United States Capitol, which is a famous and unique building, but these cases should be rare indeed.

Glare is unpleasant as well as dangerous. Anyone unsure of this need only remove lampshades from lamps in the living room to find out just how grating it can be. Many recall, in any number of film noir movies, a cheap, tawdry hotel room scene with one bare bulb dangling from the ceiling at the end of a cord. No less unpleasant is glare-producing splashy lighting from fast-food restuarants, gas stations, and used-car lots at night. This junk or trash lighting was almost always put up without thought about the consequences, and new legislation seeks to improve or replace it.

Sunset and bare bulbs are not the only examples of glare. One-upmanship calls for a careful choice of seating in an office meeting. The

25.1   United States Capitol at night

one in control selects a seat directly in front of a window, whose glare masks the subtler tones of his facial features. Good cops and bad cops alike seat the accused in the chair facing the spotlight. High beams on headlights are universally disliked by those in cars directly in front of them as well as those driving in the opposite direction facing them.

Shielding from glare and upward-shining light is a win-win situation offering satisfaction to all interests along with a great saving of energy and money. New light fixtures shield the light source above the horizontal plane and from the sides. They achieve the same light levels downward with a considerable reduction in bulb wattage and power consumption. The present waste in electricity translates directly into the consumption of coal and oil, with the attendant addition to global warming through the emission of carbon dioxide and other greenhouse gases into the atmosphere.

Eyesight deteriorates with age. As the IDA (International Dark-Sky Association) states it in one of its information sheets, "There are major concerns about good lighting for the elderly. Note that almost all recommendations about lighting levels and other parameters have been determined solely for the younger eye. The older eye is different." Age brings on the need for more light for low-contrast tasks, greater sensitivity to glare, need for uniformity in lighting, and the well-known presbyopia, in which change of focus falters without bifocals. All reinforce IDA's position on the need for quality nighttime lighting and a better nighttime environment.

Of great importance to our aging population is the issue of glare. Elderly drivers have a harder time with glare and high-light contrast than others. For this reason, the Massachusetts Medical Society is cosponsoring a bill in that state strictly enforcing limits to glare and skyglow, in view of the serious safety hazard for those with cataracts or aging eyesight, whose mobility at night is now severely limited by the presence of glare. Other groups representing older people nationwide are soon likely to do the same. With greater sensitivity to glare, drivers will need more than ever bright light sources being placed out of direct view or shielded so direct glare is avoided. The only thing we can do to decrease or avoid disability glare is to screen the glaring light from reaching the eye.

Again the IDA:

> Adaptation control is important. The time required to adapt to changing light levels is much longer for the aged. This means going from bright areas to dark ones, and the reverse, can be a real problem. Good visibility is difficult when moving between spaces lighted at very different light levels. The adaptation time is longer when going from bright to dark than the reverse. One example seen frequently these days is in going from an overlit service station back out onto the street or highway. Ironically, the excessive lighting is "justified" in terms of safety rather than marketing. Similar differences can be seen when going from overilluminated entryways to walkways or parking lots. The solution is not to raise lighting levels in all the darker areas (an impossible task) but to use rational levels in the brighter areas. Large differences in light level should be minimized in transition areas, such as building entrances, parking lots, walkways, and streets. While differences in light level are important

in any lighting design, they are of paramount importance for the elderly. In addition, lighting should not be directed toward the viewer's eyes so as to cause glare and veiling luminance.

Science can be thought to start with quantitative measurement. This applies to glare, which is not well measured or numerically defined at present. A bright source amid a darkened field is the usual glare situation, but how bright is the light source and how large is the field of view? The ratio between the brightness of the source and the brightness of the background is a first step in the quantification of glare. But in the end, "quality, not quantity, of lighting is what counts. By quality one means: freedom from glare, freedom from veiling reflections, freedom from flicker, smooth transitions, and appropriate spectral distribution. Quality night lighting takes into account the strengths and weaknesses of human vision." Above all, quality lighting does not tolerate glare.

✳

26

The Endangered Heavens

Folly—pursuing policies injurious to self-interest while being
advised against them—is nothing new; it has plagued governments
since their inception.

Barbara W. Tuchman, *The March of Folly*

The George Washington Bridge is very big; with fourteen lanes on two
levels, it is the busiest suspension bridge in the world. A few years ago,
the Port Authority of New York and New Jersey had the idea of illuminating its two 600-foot towers from within. In due course 688 1,000-watt
metal-halide fixtures and 72 150-watt lamps were installed and in the
summer of 2000 they were turned on for the first time.

A suspension bridge combines beauty with utility as do few other
structures of any kind, and the George Washington Bridge is one of the
most impressive. Its bright lights have been turned on for only a few of
the year's holidays, but if they were on every night they would pose a
double threat in addition to glare and light trespass. As already mentioned, evidence has shown that migratory birds are killed in very large
numbers by bright lights, especially when mounted on towers. Birds follow insects into the glare, and many of them, especially songbirds that
use lights to navigate, become confused and fly to the point of collision
or collapse.

A second and even more threatening potential for harm comes from
the recently confirmed disruption of melatonin (a hormone of the pineal

gland that humans secrete at night) from bright light in the nighttime hours. The circadian rhythm of day and night, shared by all life-forms, is now known to be very beneficial; melatonin is aided by this rhythm and plays a role in reducing the incidence of cancer, particularly breast cancer. If nighttime lighting disrupts the normal secretion of melatonin, the rate of cancer is likely to rise. People living near a bright light source may be victims of it in a way not known just a few years ago. This double jeopardy may well give pause to many gaudy lighting projects in the near future. An example of nighttime beauty can and too often is an example of hubris.

Hubris, wanton insolence or arrogance resulting from excessive pride or from passion, is behind much human behavior. Hubris is directed toward many kinds of ambitions and subjects. For example, toward nature it took an almost hostile attitude more prevalent a century ago than today. Typical of this reduction is our emerging attitude toward one of the unique spots on the planet. The Galápagos Islands straddle the equator in the Pacific Ocean to the west of Ecuador. They are a group of volcanic islands, similar to Hawaii, and in this sense they are oceanic islands as opposed to continental islands, such as, for example, the West Indies, which have been associated with the continental shelf in the recent geological past, as parts of continents. Like Hawaii, the Galápagos archipelago formed over a hot spot spewing lava from the ocean floor; as plate tectonics carry a plate over the hot spot, new islands emerge from its residue. The largest island in the State of Hawaii is the big island, also called Hawaii, and the hot spot is now located near Kilauea, the active volcano on its southeastern rim. In time it will move offshore as the island moves northwest of the hot spot and a new island will form to its southeast. So it is with the Galápagos, except that the next island will form to the west.

The traveler to the Tropics quickly becomes aware that twilight is of short duration; night comes quickly near the equator. This is because the setting Sun intercepts the western horizon at a steep angle, much steeper than is usually the case at midlatitudes. Absolute darkness, or at least the absence of any discernible residual twilight from sunlight, comes quickly and very completely. The night sky is so dark in the Galápagos—there's only a handful of streetlights anywhere on the islands—I was able to count the streetlights over one of its few small villages and found just twenty-three including one that had burned out. The sky on a moonless night is pitch-dark; the winter Milky Way stands out with great brilliance and the Magellanic Clouds are clearly visible. These are two small galaxies close to our own and satellites of it; in a dark sky they resemble nothing so much as two pieces of the Milky Way broken off and well separated from it. They were first reported by Magellan and his crew during their cruise around the world in 1521 and lie in the extreme southern sky and thus cannot be seen much north of the equator. The sky is absolutely filled with stars.

By far and away the most famous visitor to these islands was Charles Darwin, born on the same day as Abraham Lincoln, February 12, 1809. In 1835 the young Darwin visited the Galápagos Islands in the course of his voyage on the *Beagle,* and it was here as much as any other spot that the theory of natural selection and evolution first formed in his thinking, though it crystallized much later with his publication of *The Origin of Species* in 1859.

Not long after Darwin's visit, Herman Melville happened by the same Galápagos Islands. Melville was one who saw the dark side of both man and nature, as his famous portrayal of the struggle between an embittered sea captain and a maliciously evil whale revealed. He described the islands fairly but then drew the comparison with Dante's *Inferno.* "It is to be doubted whether any spot of earth can, in desolateness, furnish a parallel to this group," he writes, claiming of them an emphatic uninhabitableness, a special curse of changelessness, and adding, incorrectly, that rain never falls on these blighted islands. "No voice, no low, no howl is heard; the chief sound of life here is a (reptilian) hiss."

Melville was racked with the hubristic attitude common in his day, shared by our whole society, that the Earth and nature were there for us to manipulate as our leaders in business and industry chose. Think of Melville's years in the mid-nineteenth century, at a time when man's fantastic progress with machines and communications brought the world together with instantaneous contact for the first time in history. In the early twentieth century, the world witnessed the loss of the unsinkable

Titanic and the Hindenburg disaster, signaling the fact that man cannot control nature or should not attempt to control nature. The *Titanic* disaster, followed by the desolation of the First World War, cast a pall over society; for a time we lost much of our boundless optimism, but with it some of our infernal conceit.

How different is this archipelago as seen through the eyes and pen of the contemporary author Annie Dillard. Contrast her account with Melville's impressions of the Galápagos Islands written in her *Teaching a Stone to Talk*. In it she reports seeing them as the "fascinating world . . . one to be treasured and saved and relished . . . natural delights and wonders." She "knelt on a plain of lava boulders . . . stroking a giant tortoise's neck." In accord with my own experiences there, she writes of the tameness of the animals. They show no fear of humans. The arrogance rising in Melville's account of these worthless islands and their manipulation at will by our kind gives way to a love of nature and this unique environment, treating it as the world park it has every right to be.

But arrogance may be making a comeback in these economically favorable times. America's attitude toward its natural environment is once again hardening; even many conscientious folk are opting for SUVs, pickups, and other large vehicles that guzzle just as much gas as did the chrome behemoths of the fifties. We charge about on wasteful internal combustion gadgets—snowmobiles, dune buggies, and powerboats, tempting our way back into oil shortages, greenhouse gas emissions, smog, and asthma.

Nowhere is this more evident than in our present callous treatment of the sky. Our fumes fill the air with unhealthy aerosols and then we illuminate them from below. The latest fancy involves the placing of bright floodlights on our bridges, towers, and church steeples, shining upward of course. We have cited the George Washington Bridge as an example of this excess; let us consider one more, this time a tower. The Space Needle, Seattle's 600-foot pylon, was erected as a part of a 1962 exhibition. In 1999, the city fathers were persuaded to install spotlights of great intensity shining mostly upward. Most local citizens live below the top of the tower, so the full effect of this glare is reserved for a few pilots, an occasional astronaut, and birds. Birds follow insects into the glare and many migratory birds, especially songbirds that use lights and stars to navigate, fly until they die from exhaustion or from colliding with a tall structure. In his recent book on migratory birds, *Living on the Wind,* Scott Weidensaul notes that towers are harmful to flocks of birds en route from one spot to another. He writes, "Towers themselves are dangerous enough, but when you add lights, the situation gets considerably worse. In bad weather, lights disorient night-flying songbirds, which rely on subtle glimmers from the stars and the shadowy tapestry of the landscape for navigation. Under low clouds or in fog, lights overwhelm these cues, forcing the birds to circle lighted towers and buildings like moths around a flame, eventually ramming themselves to death or collapsing from exhaustion . . . lights lure millions of migrants to their deaths."

Contrast this Seattle structure with a Christmas tree. This 600-foot-tall metallic structure could have been designed to appear like a holiday spectacle draped with small nonglaring tracer lights shielded from above, white or colored as the case may be, along the sides of the structure outlining its shape. The citizens might in that case see a delightful addition to the Seattle skyline. Interference with the stars would be minimal, and the birds would not die by the thousands or seek their migratory routes elsewhere, leaving the city to a possible overabundance of insects. Instead of a return of Rachel Carson's silent spring, the city would enjoy an attractive and inoffensive spectacle. But Seattle has followed that most garish of nighttime displays, Las Vegas. Hubris again enters in a big way whenever one wants one's city to be visible from far out in space, maybe even from the Moon.

✳

Godel's Theorem and Other Science Esoterica

Godel's theorem, the theorem that in a formal logical system incorporating the properties of the natural numbers, there exists at least one formula that can be neither proved nor disproved within the system.

Webster's College Dictionary

In the last few years, it has become fashionable to disparage science from the academic left. Charges have been leveled such as the one that categorized the central figures in the rise of science and, to a lesser extent, European civilization in general as DWEMs (Dead White European Males) and therefore no longer worthy of study or applicable in this postmodernist society.

Whereas almost every civilization has made contributions to our sum of knowledge, they did not do so equally. To ignore one of them in favor of another is to misunderstand both. *Guns, Germs, and Steel: The Fates of Human Societies,* the recent bestseller by Jared Diamond, presents a much more balanced and benign view of the reasons for the early dominance of Europe (along with eastern Asia). No assertion is made that those who lived there were smarter than those who did not. For the most part their predominance was an accident of geography.

In a grand send-up of the postmodernist point of view, Alan Sokal wrote, submitted, and published an article to a fashionable American cultural studies journal in 1996. The satirical essay was entitled "Transgressing the Boundaries: Toward a Transformative Hermeneutics of Quantum Gravity." It is a parody of the type of work, common in the last few years, that expounds on what has become known as the postmodernist point of view. Among its absurdities is the contention that such physical staples as the ratio, π, and Newton's gravitational constant, G, are no longer to be thought of as constant and enduring, relevant, or part of an external world, independent of humanity. Summarizing the parody in their book *Fashionable Nonsense,* Sokal and Jean Bricmont show that this and other critiques of science go beyond attacks of its worst aspects (militarism, sexism, etc.) by others, into a condemnation of science as an "intellectual endeavor aimed at a rational understanding of the world" by those who do not themselves comprehend science.

Godel's theorem, Heisenberg's uncertainty principle and quantum physics in general, Mandelbrot sets and fractals, chaos theory, the big bang, the double helix, and the special and general relativity theories of Einstein are all among current scientific esoterica. Some by their very nature are recondite matters at best, but this is no argument against their validity. They are often used and thoroughly misunderstood by those who detail the argument against rational thought of the post-Renaissance enlightenment. In these entanglements is the postmodern left so far from the know-nothing attitude of the far right? Many of the latter cling to their moribund beliefs in the Bermuda Triangle, astrology, Atlantis as a midocean continent bearing an advanced civilization, the Full Moon effect, the Shroud of Turin as clothing worn by Christ, along with creationism and religious intolerance, which all go to fulfill a more traditional distrust of science.

Fortunately, not all antiscientific peccadilloes of the left and the right survive; two that seemed credible a generation or so ago, when two authors proposed theories along these lines, have become passé in present times. The authors are Immanuel Velikovsky and Erich von Däniken. Von Däniken's thesis, detailed in several books, purports that earlier societies could not and did not build such wonders as the great pyramids without the help and instruction of advanced aliens from beyond our solar system. Not a scintilla of proof accompanies this far-fetched notion. We built those things, and he sold our species short.

The case of Immanuel Velikovsky is even more bizarre. He developed a theory that the flight from Egypt and Pharaoh by the people of Israel over 3,000 years ago as recounted in the Bible was accompanied by a series of solar system–sized catastrophes. Velikovsky held that a mass that was later to become Venus tore out of Jupiter at about that time, that is, within the period of recorded history, to pass closely by the Earth on its way into its present orbit. In the course of this near miss, the proto-Venus stopped the Earth's rotation for a while, thus fulfilling Joshua's legendary commandment to the Sun to stand still, which Velikovsky accepted as partial proof of his case.

His first of several books, *Worlds in Collision,* was published in April 1950. This came about just a few weeks after Senator Joseph McCarthy first asserted that 57 or 205 or some other number of employees in the State Department were security risks if not outright communists serving the atheistic government of the Kremlin. The two events are neither part of a causal relationship nor a conspiracy, but their coeval origin in a period of national distrust and paranoia does suggest overlapping causes.

Velikovsky's publisher, Macmillan, was one of the most respected in the nonfiction world. Many scientists rightly criticized the theory from a number of directions, but a few of them made a critical mistake; they pressured (as best they could) Macmillan to dissociate itself from Velikovsky and his views. This only heightened widespread interest in the old curmudgeon, as any forethought might have predicted. (Here is an error that astronomers and most in associated fields are not about to make again anytime soon.) This example of high-handed behavior was particularly devastating in that time of antiscientific attitudes.

Sputnik in 1957 changed all that, but now we have again descended into a period of antiscience. Science is getting squeezed between the postmodernism of the left and the more serious mumpsimus (complete with a choir of angels) from the right. We see Kansas and other states pass laws favoring creationism over evolution in the classroom (the

Kansas statute itself has fortunately been repealed). We see books such as John Horgan's *The End of Science* pick up the theme pronounced but then shrugged off by physicists a century ago, right after the discovery of relativity and quantum physics, that the big questions in the physical sciences have all been answered as much as they can be answered. Horgan quotes reputable scientists as being discouraged about their field of research, but his sensational account misses the point. After a time in which the science enterprise has seen a period of unparalleled growth, the discouragement is not with the research field per se, but with the dwindling support for funding of science by society through government, along with the fallout of science through the dumbing of America, particularly in our schools. Not by chance does the fitness craze demand "scientific breakthroughs" each day on the televised news programs with little or no corroboration.

Through proper diet and exercise, you too can cure cancer, migraines, and epilepsy at will, providing you realize that today's panaceas may differ from the nostrums promoted yesterday. Some of this nonsense is the direct result of a trend in recent years that merges the mission of academic medicine into that of corporations that have replaced independent and unbiased sources. According to columnist Ellen Goodman, "some businesses try to repress or change the results of research they funded but didn't like." Similarly, the tobacco industry promotes research that finds no link between smoking and lung cancer. Astronomy is blessed in having little such threat to the sky from profiteers unless they launch advertising satellites whose glare would blot out the stars.

The end of science has become a fashionable catchphrase in the last decade, and Horgan and others feel that the big questions in the physical sciences have been answered as much as they ever can be answered. Gerald Holton in *Science and Anti-Science* is one who disputes this dour outlook with a lively and balanced approach and, like Horgan, traces it back to the historical theories of Oswald Spengler (in *The Decline of the West,* published more than 70 years ago). I feel more comfortable with Holton's position and with the more prosaic outlook offered by Derek Price (*Science Since Babylon*), who demonstrates that the finite resources of society are eventually bound to restrict the growth of science. It can no longer continue at an exponential rate but will approach saturation, as must the world population in time. Both Spengler and Price offer our current break between centuries as the approximate turnover time, from different premises.

Those who perceive the demise of science with relief may find encouragement from Saint Augustine who, in *The Confessions,* preached against "the disease of curiosity," comparing it to the lust of the flesh. He notes that the "curiosity and the skepticism it engenders are not popular attributes at either end of the political spectrum or among religious groups whose truth is received rather than searched for."

Augustine further cautions that "one who can measure the heavens, number the stars, and balance the elements is no more pleasing to God than one who cannot. Scientific knowledge is more likely to encourage pride than to lead people to God." Salvation was his goal, not material progress; science was not only superfluous to that aim, but it might even be dangerous. The increasing scientific illiteracy of today could endanger our society by moving it toward a world of this pious desolation. Whatever the case may be, those educated in science are better able to deal with broad problems such as global warming than those who are not.

I, for one, shall continue to look upon the heavens and their wondrous fill of stars. They seem to me to give rise to a boundless curiosity, to know how big they are, how far they are, and how they got there at all. To require one or another form of a God or gods seems to me to beg the question. How the night sky got there is more to the point than who put it there and with what motive.

But in any event, and despite assaults from left and right and wasteful outdoor lighting and glare, the night sky remains there for us all to experience and appreciate as our birthright. The stars are together the most compelling sign we have that something in this universe is greater than pettiness at any level.

Let's finish with a well-known statement by Ralph Waldo Emerson that appears in the first chapter of his treatise, *Nature.* There he stated: "To go into solitude, a man needs to retire as much from his chamber as from society. I am not solitary whilst I read and write, though nobody is with me. But if a man would be alone, let him look at the stars. The rays that come from those heavenly worlds, will separate between him and what he touches. One might think the atmosphere was made transparent with this design, to give man, in the heavenly bodies, the perpetual presence of the sublime. Seen in the streets of cities, how great they are! If the stars should appear one night in a thousand years, how would men believe and adore, and preserve for many generations the remembrance of the city of God."

Although Emerson posed this as an imagined fancy, it has now come true for millions of people in all lands. We have imposed a curtain of uplight and glare that has rendered the stars nearly invisible for most citizens almost all of the time; only when we escape the glare of civilization, be it but once in a year or in a lifetime, can we experience and appreciate what he and others have tried to express. That curtain must be drawn aside in order to restore our truest view of the sublime.

*

Appendix I

There! See the line of lights,
                A chain of stars down either side the street—
Why can't you lift the chain and give it to me,
A necklace for my throat?

                                            Sara Teasdale

During the fall semester of 1996 I was appointed a fellow at the Wesleyan Center for the Humanities and, as fellows are expected to do, I taught a course as a part of its program. Natural scientists were not customarily named to the faculty of the Center, and I sought to offer a course blending scientific and social content. The course was centered on light pollution and night-sky brightness, and their physical and social implications. It was the first course on light pollution taught anywhere for college credit. The theme, as befits the title, was that the perception of outdoor lighting is moving from the simplistic "more is always better" syndrome to a paradigm in which much of it is seen to be excessive and wasteful, perception being always dominant over reality here wherever the two differ. The information sheets of the International Dark-Sky Association formed the must useful textual material of any source. In 1996, Wesleyan University had become a different place from the campus of a decade earlier.

From 1976 to 1990, forty-six incidents of sexual assault or sexual contact, usually made by a male on a female, had been reported by the

Office of Public Safety of the university. (Reports of corresponding violence of other kinds have been very low.) The events have been tabulated in Table A.1 after division into the following groups: indoor and outdoor incidents, daytime and nighttime incidents, and, for the nighttime group, those that occurred in illuminated and dark locations.

The first row of data lists the total number divided into three periods of about 5 years each. This is followed by the data divided into events occurring indoors and outdoors and those for which the location is unknown. The next group of three rows divides the outdoor incidents into daytime and nighttime events and those for which the time has not been divulged. The final set of data divides the nighttime outdoor events into those occurring in brightly illuminated areas (areas in which the level of illumination was much brighter than 1 foot-candle) and dark areas, with one incident occurring in an uncertain level of illumination.

Although the summary data depend extensively on the information available in the memos, some further information has been developed concerning the outdoor incidents occurring at night and for those for which the time of day was not provided. Most of the sites were visited by astronomers at night immediately after the incident in order to measure the lighting level at the site. The division into bright and dark in the final section of the table reflects this additional information. In addition to the incidents known to have happened at night, the thirteen of unknown time were found to be divided as follows: six happened in brightly lit areas, one in an unlit area, and for six no information about the area is available.

Five of the forty-six events occurred in locations well removed from the campus and do not lie within the region over which Wesleyan and its lighting policies have an influence. All of these took place outdoors, one in the daytime, three at night, and one at an unknown time. If these are removed from the numbers in the table, the data would favor indoor events more noticeably. A bias may be present here in the sense that off-campus events occurring indoors are less likely to be reported to the Public Safety Office than those occurring outdoors. Table A.1 shows a trend away from outdoor incidents occurring at night and at unknown times (which dominate the period from 1976 to 1980) toward indoor and daytime events in the

TABLE A.1

Distribution of Incidents of Personal Molestation
On or Near the Campus of Wesleyan University

	1976–80	1981–85	1986–90	Total
Total	18	16	12	46
Indoor	4	6	4	14
Outdoor	13	7	8	28
Unknown	1	3	0	4
Outdoor				
Daytime	0	1	4	5
Nighttime	4	2	3	9
Unknown	9	4	1	14
Outdoor, Nighttime				
Bright	3	0	2	5
Dark	1	2	0	3
Unknown	0	0	1	1

years since 1980. As noted above, the Wesleyan campus was most brightly illuminated during the period from 1976 to 1980, after which time some reduction in the sky brightness came about through shielding and other measures in response to the requests of the astronomy department and conservation groups. This earliest period contains the majority of events for which one or more aspects remain unknown; since 1980, the number that cannot be fully classified is very small.

From this sample, we can conclude that at least two and at most ten of the forty-six incidents occurred in darkened areas at night. If we include the Wesleyan campus area only, the range extends from a minimum of zero to a maximum of six. The maxima require that none of the uncertain cases occurred indoors or in the daytime or under a streetlight, and the minima require that one or more of these circumstances applies to each of the uncertain cases. It would appear that at Wesleyan there is no evidence that outdoor lighting forms a deterrent to crimes of sexual assault and contact. At first glance it appears that the outdoor illumination level and the frequency of incidents are correlated, suggesting that brighter nighttime conditions actually promote crime of the kind under consideration here. However, the sample size is too small for this conclusion to be made with confidence. It does strongly indicate the absence of a negative correlation, long believed by the public at large. It supports similar findings at a national level by Tien et al. (1979), among others, who draw the distinction between being safe and feeling safe. Excessive outdoor lighting levels do promote the latter condition, but proven safety measures would appear preferable. Further study is needed to substantiate the role of lighting, if any, in the incidence of crime. But public perceptions deserve to be examined critically and, if found wanting, publicized and corrected.

*

Appendix II

Johannes Kepler's third law of planetary motion reveals an instance of how science proceeds. The law states that

$$P^2 = a^3$$

which means that the period of a planet in years squared equals its average distance from the Sun in astronomical units cubed. For Earth, this is simply $1 \times 1 = 1 \times 1 \times 1 = 1$ by definition. For the other planets the mean distance, a, can be found if we just observe its period. Thus, for a hypothetical planet at 4 AU (four times our distance from the Sun), we get $4 \times 4 \times 4 = 64 = 8 \times 8$, or an 8-year period. This is not as straightforward as it may sound because we observe the other planets not from a stationary platform but from one of the orbiting worlds. The true period cannot be directly observed, but must be inferred.

Kepler devised a simple rule that determines the true period from the length of our year and the apparent period of another planet. Since the apparent periods are not constant but vary by a few days from one to another, many years of observations must be obtained before a precise value is possible. But another more significant reason for error in the long run is that Kepler did not work with the complete equation, which he could not have known. It remained for Isaac Newton to discern the complete picture; here, the masses of the planet and the Sun added together form a fundamental feature of the orbital period of the one about the other. The correct expression as Newton derived it is that P^2 times the sum of the two masses is what is equal to a^3; that is, M(Sun) + M(planet) times $P^2 = a^3$, or in its common form

$$M(Sun) + M(planet) = a^3/P^2$$

Kepler missed the role of the masses in part because the Sun is so much heavier than any of the planets, being a third of a million times the mass of Earth, and even a thousand times the mass of Jupiter, the largest planet. With Newton's correction, or rather his generalization, the proper term is a little different from unity, as Table A.2 shows. In it the mass given is in terms of the Sun's mass alone. With no mention of mass, Kepler used the equivalent of the Sun alone, unity in every case. As seen from the third column, Newton, by adding each planet in its turn, used slightly greater values. In Kepler's day observations were insufficiently accurate to distinguish between 1.000 and nearly 1.001, the two models for Jupiter's motion; the larger calls for a slightly shorter period for a given distance. Oddly enough, it was no closer to resolution in Newton's time either, but his great insight provided the clue that the masses are important components in his gravity-dominated cosmos. Once the great success of Newton's model was established, the sought-for difference was revealed in the observations.

This is the full expression for Kepler's third or harmonic law, and it has proved to be a bonanza for the astrophysicist. The most important parameter of a star or a planet is its mass. Know the mass of an object and you have a fairly realistic picture of its life history, what kind of star it has been or will become, as well as its present state. As telescopic observations improved throughout the eighteenth century, astronomers found pairs of stars in the sky with much greater frequency than would be expected from a random distribution. Sir William Herschel, the discoverer of Uranus in 1781, observed that some of the closest pairs of double

TABLE A.2

The Differences for the Planets
When Their Masses Are Considered

Planet	Kepler's Mass	Newton's Mass
Mercury	1.000	1.00000017
Venus	1.000	1.00000247
Earth	1.000	1.00000303
Mars	1.000	1.00000032
Jupiter	1.000	1.00096
Saturn	1.000	1.00029

stars showed orbital motion about each other over the years, and from the sizes and periods of the orbits, the masses could be directly determined.

The universality of this and other of Newton's conclusions was quickly established. The masses of the planets are revealed from the periods and mean distances of the moons that orbit them (all but Mercury and Venus). And a straightforward extension of Newton's laws derived their masses, and that of the Moon as well, from their gravitational perturbations on their neighbors. What better proof that his "universal" law of gravitation is truly universal; it unites the solar system and the greater galactic system into a common physical reality.

＊

Glossary

Aerosol: A suspension of a particle or particles in the air.

Airglow: The light produced in the Earth's atmosphere, mostly emitted by the upper atmosphere.

Air Mass: The amount of air between a star or planet and an observer. By definition, the air between observer and the zenith is equal to 1.

Albedo: The percentage of light falling on a body or surface that is reflected directly back into space.

Analemma: The figure-eight shape made by the Sun in a year as recorded at one particular time of day.

Aphelion: The point that is farthest from the Sun in the orbit of an object going around it.

Apparent Solar Time: A measure of time based on the daily motion of the Sun as we see it. It is the time recorded by a sundial and is not uniform.

Apsides: The two points on an elliptical orbit lying closest and farthest from the center of mass; for a planet they are known as perihelion and aphelion

Astrometry: The branch of astronomy that deals with positions and motions of celestial objects.

Astronomical Twilight: The period of time during which the Sun lies between 12° and 18° below the true horizon, or, more properly, when it is between 102° and 108° from the zenith.

Astronomical Unit: The mean distance between the Earth and the Sun. It is defined as 149,597,870.66 kilometers or about 92,955,807 miles.

Barycenter: The center of mass of two or a system of objects moving under the influence of their mutual gravity.

Bioluminescence: The phosphorescent glow from a living organism, usually in the sea.

Candela: The standard unit of luminous intensity; 1 candela equals 1 lumen per steradian.

Candlepower: The luminous intensity expressed in candelas.

Celestial Equator: The great circle on the celestial sphere directly above the Earth's equator.

Circadian: Of or pertaining to rhythmic cycles recurring at or about 24-hour intervals.

Civil Twilight: The period of time during which the Sun lies between 0° and 6° below the true horizon, or, more properly, when it is between 90° 50′ and 96° from the zenith since it begins (or ends) when the refracted upper limb of the Sun reaches 90° below the zenith.

Colure: Any of the two equinoxes or the two solstices or any great circle passing through either pair and the celestial poles.

Conic Section: Any circle, ellipse, parabola, or hyperbola.

Conjunction: An alignment of two celestial objects with the Earth so that they appear near each other in the sky.

Coriolis Effect: An apparent effect or force upon a moving object seen in a rotating frame of reference.

Creationism: The doctrine that the true story of the creation of the universe is recounted in the Bible.

Declination: The angular distance of an object or point from the celestial equator measured along the great circle passing through the point and the celestial pole. It is one of the coordinates used to define position on the celestial sphere and is analogous to latitude on the Earth.

Eccentricity: A measure of the shape of a conic section; it is 0 for a circle, 0 to 1 for an ellipse, 1 for a parabola, and greater than 1 for a hyperbola.

Eclipse: An event where one body passes in front of another, covering it, or the passage of a body into the shadow of another. Solar eclipses are examples of the former, and lunar eclipses of the latter.

Ecliptic: The apparent path of the Sun in the sky, or the plane passing through this path and the Earth.

Equation of Time: The apparent solar time minus the mean solar time. The two times can vary by as much as 15 minutes at specific times of the year.

Equator: The great circle on the surface of the Earth or other object passing through its center and perpendicular to its axis of rotation.

Equinox: Either of the two points where the celestial equator intersects the ecliptic.

Extinction: The reduction in light intensity with angular distance from the zenith, measured in magnitude reduction per air mass.

Flux: The rate of flow of light past a particular point.

Foot-candle: The unit of illuminance from a source at a distance of 1 foot.

Fovea: A small region surrounding the center of the retina containing only cones but no rods, allowing particularly acute vision.

Fractal: A structure with a shape repeated over all scales of measurement.

Full-Cutoff (FCO) Shielding: Shielding around an outdoor light fixture that prevents any direct light from the bulb shining above the horizontal plane.

Galaxy: A group of many stars held together by mutual gravitation and with a distinct identity separating it from other galaxies.

Geocentric Model: A model of the solar system that places a stationary Earth at the center of motion.

Glare: The sensation produced by light within the visual field sufficiently greater than the luminance to which the eyes are adapted to cause annoyance, discomfort, or loss of visibility.

Greenhouse Effect: The internal heating of an atmosphere due to its opacity to infrared radiation. The effect is similar to a greenhouse in which the glass is transparent to radiation in the visual region from the Sun but opaque to the infrared reradiation, heating up the interior as a result.

Heliocentric Model: A model of the solar system that places the Sun at or near the center of motion of the solar system.

High-pressure Sodium (HPS): The high-pressure sodium vapor lamp is a high-intensity discharge lamp that shines across the red, orange, and yellow portion of the visual spectrum. It appears orange to orange-pink, and is the commonest streetlight in most places.

Horizon: In astronomy, the horizon is assumed to consist of a single straight line separating the surface from the sky; it is approximated by the horizon visible from a rowboat at sea.

Hubble Constant: The rate of expansion of the universe with time, appearing as the recession of galaxies from each other at a velocity proportional to their separation.

Illuminance: The density of luminous flux incident on a surface; it is the quotient of the luminous flux by the surface area. It is commonly measured in foot-candles or lux and can be thought of as the brightness received at a point. Compare with luminance.

Light Pollution: Any adverse effect of man-made light. Often the term is used to denote urban sky glow.

Light Trespass: Light spilling onto a neighboring property whose owner objects to its presence.

Light-year: The distance traveled by a light beam in a vacuum in one year, at the rate of 299,792.458 kilometers or about 186,000 miles per second. This amounts to about 6 trillion miles or 9.5 trillion kilometers.

Low-pressure Sodium (LPS): The low-pressure sodium vapor lamp is a discharge lamp that shines at only one color, a narrow yellow-orange portion of the visual spectrum. It is less costly to operate at an equivalent brightness than the high-pressure sodium lamp, but has poor color rendition.

Lumen: A unit of luminous flux.

Luminaire: A complete lighting unit consisting of one or more lamps together with the parts designed to distribute the light and to position the lamp and connect it to the power supply.

Luminance: The amount of light propagated by a source as opposed by illuminance, light received. This is an approximate definition; the exact definition can be found in any lighting reference. Luminance is commonly measured in candela per square foot or square meter.

Lunation: The length of time between successive appearances of the same lunar phase (thus a complete cycle of the phases of the Moon). Also known as the synodic month, this period averages 29½ days.

Lunisolar Calendar: A calendar, such as the Jewish calendar, that is based on motions of the Moon as well as the Sun.

Lux: The unit of illuminance from a source at a distance of 1 meter or 3.28 feet. One foot-candle is equal to 10.76 lux.

Mean Solar Time: A measure of time based on the daily rotation of the Earth, assumed to pass at a constant rate. It is the time registered by a properly running clock.

Mercury Lamp: A high-intensity discharge lamp in which most of the light is produced by radiation from mercury. The light appears bluish and is in declining use because it conserves less energy than sodium vapor lamps.

Mesopic Vision: Vision with fully adapted eyes at luminance conditions between those of photopic and scotopic vision.

Metal Halide Lamp: A high-intensity discharge lamp in which most of the light is produced by radiation of metal halides. It commonly appears white or bluish-white in color.

Milky Way: The galaxy in which the solar system and all naked-eye stars are located. It appears as a hazy band of light circling the sky. The word *galaxy* is generally capitalized when referring to our galaxy, the Milky Way.

Mumpsimus: A persistent belief in the face of conclusive contradictory evidence, or a person who obstinately clings to such a belief.

Nanolambert: One billionth (10^{-9}) of a lambert, a billionth of a unit of luminance equal to $1/\pi$ candela per square centimeter.

Nanometer: One billionth (10^{-9}) of a meter.

Nautical Twilight: The period of time during which the Sun lies between 6° and 12° below the true horizon, or, more properly, when it is between 96° and 102° from the zenith.

Neap Tide: The minimum amplitude between high and low tide that occurs when the Moon is near first- or last-quarter phase and thus is at a right angle with the Sun and the two gravitations offset each other, with the Moon producing the stronger effect.

North/South Celestial Pole: The point in the sky directly above the North/South Pole of the Earth.

North/South Ecliptic Pole: The point in the sky 90° north or south of the plane of the ecliptic.

Nyctophobia: An unnatural fear of darkness.

Oblateness: Ellipticity; the ratio of the difference between the polar and equatorial diameters of a planet divided by the polar diameter.

Occam's (Ockham's) Razor: The principle in science that assumptions used to explain something must not be multiplied beyond necessity; thus, the simplest of several hypotheses is the best.

Occultation: The passage of a celestial object across a smaller one, causing the latter to be hidden from view.

Olbers's Paradox: The dark night sky appears to conflict with a universe of infinite age filled with stars, in which the sky should be a continuous blaze of light because every line of sight would ultimately encounter a star. Yet the sky is dark because in the finite time since the big bang the universe cannot have gotten filled with stars and starlight.

Opposition: The position of a superior planet (a planet farther from the Sun than the Earth) when it is opposite the Sun in the sky.

Parallax: The change in the relative positions in angular measure of objects viewed from different places. In the case of nearby stars appearing to shift due to the Earth's orbital motion, it is called trigonometric or heliocentric parallax and is a direct measure of the star's distance.

Paranormal: Of or pertaining to events or perceptions occurring without scientific explanation.

Perihelion: The point that is closest to the Sun in the orbit of an object going around it.

Photopic Vision: Vision produced by the cones, the retinal receptors that dominate the visual response when the luminance level is high and produces the perception of color.

Precession (of the Equinoxes): The uniform motion of the rotation axis of an object such as the Earth due to external gravitational influences. It causes the Earth's axis to sweep out a cone of angular radius of about 22° to 24° in 25,800 years due to the gravitation of the Sun and the Moon on the Earth's equatorial bulge.

Proper Motion: The apparent motion of a star across the sky, measured as the angular shift in its position per year.

Pseudoscience: An unscientific or unsound discipline that attempts to pass for an alternate science, often in the service of political ambition, but does not succeed in achieving a body of testable and replicable knowledge. Compare with science.

Quintessence: The pure essence of a substance, specifically the fifth essence or element as opposed to earth, air, fire, and water, the four elements of the earthly realm. Aristotle and others supposed that the celestial realm was composed of this fifth element.

Right Ascension: One of the coordinates used to define position on the celestial sphere; it is analogous to longitude on the Earth.

Science: A system featuring cumulative growth of knowledge over time, in which useful features are retained and nonuseful features are abandoned, based on the confirmation or rejection of testable knowledge. Compare with pseudoscience.

Scotopic Vision: Vision produced by the rods, the retinal receptors that dominate the visual response when the luminance level is low. Rods are more sensitive than cones but can perceive no differences in color.

Seasonal Affective Disorder (SAD): Recurrent winter depression characterized by oversleeping or irritability and relieved by light therapy or the arrival of spring.

Sidereal: Pertaining to the stars. Sidereal motion is motion without respect to objects within the solar system.

Skyglow: The background light in the night sky that prevents it from being totally dark.

Solstice: Either of the two points on the ecliptic where the Sun reaches an extreme north or south of the celestial equator.

Spring Tide: The maximum amplitude between high and low tide that occurs when the Moon is near new or full and thus is in line with the Sun and the two gravitations reinforce each other.

Steradian: A solid angle subtending an area on the surface of a sphere equal to the square of its radius. The entire surface comprises 4π steradians.

Summer Triangle: A triangle formed by the three brightest stars near the zenith in the summer. These are Vega, Deneb, and Altair.

Syzygy: An alignment or near alignment of three celestial objects such as the Earth, the Moon, and the Sun.

Terraformation: Alteration of the surface and environment of a planet or other object in order to make it capable of supporting terrestrial life-forms.

Transit: The passage of a small object across the face of a larger one, such as Mercury or Venus across the disk of the Sun.

Wavelength: The shortest distance between two points in a wave pattern that have the same phase, or the distance between one peak and the next.

Zenith: The point directly overhead above an observer.

✳

Bibliography

Isaac Asimov, "Nightfall," in *Nightfall and Other Stories* (Ballantine, 1984). Asimov's first big success; at twenty-one he wrote and published this account of a civilization in a planet in a sextuple star system. Only once in a millennium did all six suns set as seen from one location; when this happens, the inhabitants, frightened by dark, riot and destroy their entire society.

Hanbury Brown, *The Wisdom of Science: Its Relevance to Culture and Religion* (Cambridge University Press, 1986).

Tertius Chandler, *Four Thousand Years of Urban Growth: An Historical Census* (St. David's University Press, 1987). Chandler determines the population of the world's large cities over the centuries using a number of statistical comparisons.

Bernard Cohen, *Science and the Founding Fathers* (W. W. Norton, 1997). Cohen traces how Isaac Newton and his science greatly influenced the political thought of Thomas Jefferson, Benjamin Franklin, John Adams, and James Madison, as they created their new country.

Neil F. Comins, *What If the Moon Didn't Exist? Voyages to Earths That Might Have Been* (Harper Perennial, 1995). How would conditions on Earth differ if we had no Moon, or if it were smaller than it is? A delightful collection of "what ifs" in the solar system.

D. L. Crawford, ed., *Light Pollution, Radio Interference and Space Debris* (Astronomical Society of the Pacific, 1991). The proceedings of the first international symposium on light pollution; many of its articles help define the impact of excessive outdoor lighting.

Jared Diamond, *Guns, Germs, and Steel: The Fates of Human Societies* (W. W. Norton, 1997).

Annie Dillard, *Teaching a Stone to Talk* (Perennial Library, 1982). A collection of essays that includes her description of the Galápagos Islands.

David Ewing Duncan, *Calendar* (Avon Books, 1998).

Paul Ehrlich and Anne Ehrlich, *Betrayal of Science and Reason* (Island Press, 1996). A book about the environment and its detractors; it introduces the term *brownlash,* meaning opposition to ecological progress.

Martin Gardner, *Fads and Fallacies in the Name of Science* (Dover, 1957). A classic book on pseudosciences of the 1950s.

Thomas Gilovich, *How We Know What Isn't So* (Free Press, 1993).

Paul R. Gross and Norman Levitt, *Higher Superstition: The Academic Left and Its Quarrels with Science* (Johns Hopkins Press, 1997). A description of the impacts of science on academicians and others who seek relevance outside the sciences.

Jacquetta Hawkes, *A Land* (Beacon Press, reprint edition, 1991). A superbly written account of the history of Britain, tracing connections between the geological and the human impact on that land.

Carl Hempel, *Philosophy of Natural Science* (Prentice Hall, 1966).

Gerald Holton, *Science and Anti-Science* (Harvard University Press, 1993).

John Holtz, "The Number of Stars Visible," *Sky & Telescope* 87 (May 1994): 86.

John Horgan, *The End of Science* (Addison-Wesley, 1996).

Illuminating Engineering Society of North America, RP-33-99, *Lighting for Exterior Environments,* 1999.

International Dark-Sky Association Information Sheets 1-170, 1988–2000. Together these sheets form the most complete source on light pollution and all of its ramifications.

J. E. Kaufman and J. F. Christensen, eds., *IES Lighting Ready Reference,* 2nd edition (Illuminating Engineering Society of North America, 1989).

Philip Kitcher, *Abusing Science: The Case Against Creationism* (Massachusetts Institute of Technology, 1982).

Thomas S. Kuhn, *The Copernican Revolution* (Harvard University Press, 1957).

Thomas S. Kuhn, *The Structure of Scientific Revolutions,* 2nd ed. (University of Chicago Press, 1970).

David H. Levy, *More Things in Heaven and Earth: Poets and Astronomers Read the Night Sky* (Wombat Press, 1997).

C. S. Lewis, *The Discarded Image* (Cambridge University Press, 1964).

Arnold Lieber, *The Lunar Effect* (Dell, 1978).

H. P. Lovecraft, "Supernatural Horror in Literature," in *H. P. Lovecraft's Book of Horror* (Barnes and Noble Books, 1996). Lovecraft is widely regarded as the greatest master of horror and fantasy since Poe. This essay traces fear, including the fear of darkness, through literature.

Eli Maor, *June 8, 2004: Venus in Transit* (Princeton University Press, 2000).

Michael Maunder and Patrick Moore, *Transit: When Planets Cross the Sun* (Springer, 2000).

Herman Melville, "Encantadas," in Great Short Works of Herman Melville (Perennial Library, 1969). The author's impressions of his visit to the Galápagos Islands.

Marcel Minnaert, The Nature of Light and Color in the Open Air (Dover, 1954). Retranslated and reissued as Light and Color in the Outdoors (Springer, 1993).

Paul Murdin, ed., "Control of Light Pollution: Measurements, Standards, and Practice." The Observatory 117 (1997): 10–36.

Joseph B. Murdoch, Illumination Engineering: From Edison's Lamp to the Laser (Visions Communications, 1994).

Outdoor Lighting Manual for Vermont Municipalities (Chittenden County Regional Planning Commission, 1996).

Derek J. Price, Science Since Babylon (Yale University Press, 1961).

F. E. Roach and J. L. Gordon, The Light of the Night Sky (D. Reidel Publishing Company, Dordrecht, 1973).

Carl Sagan, Billions & Billions: Thoughts on Life and Death at the Brink of the Millennium (Random House, 1997).

Carl Sagan, The Demon-Haunted World (Random House, 1995).

Nicholas Sanduleak, "The Moon Is Acquitted of Murder in Cleveland," Skeptical Inquirer (Spring 1985).

Bradley E. Schaefer, "Astronomy and the Limits of Vision." Vistas in Astronomy 36 (1993): 311–61.

Michael Shermer and Stephen Jay Gould, Why People Believe Weird Things: Pseudoscience, Superstition, and Other Confusions of Our Time (W. H. Freeman, 1998).

Alan Sokal, "Transgressing the Boundaries: Toward a Transformative Hermeneutics of Quantum Gravity," Social Text 46/47 (1996): 217–52.

Alan Sokal and Jean Bricmont, Fashionable Nonsense (Picador, 1999).

James Tien et al., Street Lighting Projects, National Evaluation Program Phase 1 Report (U.S. Department of Justice, 1979).

Neil Tyson, Universe Down to Earth (Columbia University Press, 1994).

Arthur Upgren, Night Has a Thousand Eyes: A Naked-Eye Guide to the Sky, Its Science and Lore (Perseus Books, 1998).

Albert Van Helden, Measuring the Universe: Cosmic Dimensions from Aristarchus to Halley (University of Chicago Press, 1985).

Scott Weidensaul, Living on the Wind: Across the Hemisphere with Migratory Birds (North Point Press, 1999). Includes a description of the downside of night lighting and glare on bird populations.

Arthur Weingarten, The Sky Is Falling (Grosset & Dunlap, 1977). A gripping and detailed account of the collision between a bomber and the Empire State Building.

Richard S. Westfall, Never at Rest: A Biography of Isaac Newton (Cambridge University Press, 1980). A definitive biography of the great scientist.

Clifford M. Will, Was Einstein Right? Putting General Relativity to the Test (Basic Books, 1986). Descriptions of the many ways in which relativity has been proven.

Acknowledgments

I am indebted to many who helped to make this book a reality. To my agent, Sally Rider Brady, and my editor, Erika Goldman, go my heartiest thanks for their time and assurance. I would not have finished this book without their forbearance and encouragement. I thank John Wareham for his patience and care in making the figures, Gabriele Zinn for securing their permission, and Linda Shettleworth for her care in preparing the text for publication. I received encouragement and helpful comments and suggestions from David Crawford, who read the entire manuscript and, in the course of this work, from Gail Altschwager, Liz Alvarez, Joe Caruso, Bob Crelin, Charles Cutler, David H. DeVorkin, Heinrich Eichhorn, Roy Garstang, John Griese, Cliff Haas, Dorrit Hoffleit, Philip Ianna, Erhard Konerding, Alice L. Loth, Clark Maines, A. G. Davis Philip, Lewis Robertson, Bradley Schaefer, Kenneth Seidelmann, Jurgen Stock, Amy Upgren, Edward Weis, and above all the International Dark-Sky Association and a number of its staff and members for many favors at many times. I thank Jeanette Stock for providing copies of her photograph of the total solar eclipse of February 26, 1998, that appears in Chapter 3. Finally my eternal gratitude to my wife, Joan, for many helpful suggestions and for her patience with my scribblings. Any errors in this book are mine alone.

Most of Chapter 1 first appeared in *The Snowy Egret* and later as an IDA Information Sheet. A portion of Chapter 7 first appeared in *Mercury* magazine.

*

Author's Note

Since this book was written, I have been made aware that the leatherback turtle has become endangered throughout the world. This is affirmed most succinctly by, among others, Natasha Loder in an article appearing in *Nature* (the June 1, 2000, issue) in which she states, "Numbers of the genetically distinct Pacific leatherback turtle are declining precipitously." It was my experience with a member of this species that led to the epiphany described in Chapter 1, which provided the title not only for the chapter but for the entire book as well. I concluded that account with a plea that my descendants will not find a world too bright and too crowded to behold the mother turtle's descendants repeating her act of fertility in times to come. I am deeply saddened that this prospect has been made so soon a matter of uncertainty and doubt and that, by more than one account, these creatures are in imminent danger of extinction. If this book provides any impetus in the opposite direction, it will have been more than worthy of my time and effort.

✱

About the Author

ARTHUR UPGREN, Ph.D., is a senior research scientist in the astronomy department at Yale University and the John Monroe Van Vleck Professor of Astronomy at Wesleyan University in Connecticut. He is the author of *Night Has a Thousand Eyes* and the coauthor with Jurgen Stock of *Weather: How It Works and Why It Matters*.